Praise for *Resilient Agriculture, Second*

Carefully addresses the major issues facing us in agriculi
argues that it isn't some vague notion of "technology" that will show us the way
forward but people working together and carefully stewarding the land. This is
both an old and a novel approach, and it's exactly what's needed.

— Mark Bittman, author, *Animal, Vegetable, Junk* and *How to Cook Everything*

Destined to become the gold standard for laying out the climate adaptation
task and solutions before us in the 21st century, this book is both inspiring and
practical at a time when we desperately need both.

— Anne Waple, Founder and CEO, Earth's Next Chapter

With grounded hope and urgency, Laura breaks silence and reshapes social
justice language to explore issues of land and food in our changing climate.

— Meta Commerse, Founder & CEO, Story Medicine Worldwide

What Lengnick is speaking of is what the UN has been calling "eco-agriculture,"
a suite of tools and practices that not only provide greater food security but can,
scaled quickly enough, undo the worst of the Fossil Age's climate karma.

— Albert Bates, author, *BURN: Using Fire to Cool the Earth*

Wonderfully concise, practical, and beautifully written compendium of how to
deal with climate change's impacts on agriculture. This book should be on the
shelf of every farmer in America and abroad.

— Dr. Sally Goerner, research director, Edinburgh University's Planetary Health Lab

A manifesto that is one part academic exploration, one part practitioner guide,
and one part storytelling adventure that deepens our understanding of what
resilient agriculture is and what it can be. A guidepost for building a better and
more resilient food system that builds off of, and respects the foundation laid
by, local and Indigenous knowledge.

— Dr. Gabrielle Roesch-McNally, director, Women for the Land, American Farmland Trust

This is a must read for anyone working in the resilience/adaptation field or
contemplating doing so.

—Jim Fox, senior resilience associate, NEMAC+FernLeaf

Chock-full of wisdom from farmers and ranchers on the front lines of climate change. Laura Lengnick puts their stories into context and provides the path forward to building resilient food systems.

— Teresa Opheim, director, Climate Land Leaders (MN)

Everyone who works a food-related job, or who just cares about what and whether our children and grandchildren will eat, should acquaint themselves with this path-breaking, delightfully informative book.

— Richard Heinberg, senior fellow, Post Carbon Institute,
author, *Power: Limits and Prospects for Human Survival*

Highlights the bravest among us—modern day, small and medium scale, adaptability-driven farmers who are overcoming immense obstacles and staying ahead of change on every level. We have the great good fortune to learn from them and in turn learn about cultural healing and repair.

— Lee Warren, founder, Reclaiming Wisdom

An elegant and expansive exploration of food and farming in an unprecedented era of ecological violence. Lengnick compellingly offers us resilience thinking as a vast looking glass that can allow us to better grasp the complexity of the food system, nourish more holistic solutions, and stretch our socio-political imaginations.

— Daniel Macmillen Voskoboynik, author, *The Memory We Could Be*

A loving prod to the sustainable agriculture movement to go beyond sustainability to the transformation of food and farming into a commons where community-based wealth truly provides for human need.

— Elizabeth Henderson, farmer, writer

This is a pivotal book for all agricultural universities and farmers that recognize the urgent need to move beyond industrial agricultural practices.

— Dr. James Gruber, professor emeritus, Antioch University, author, *Building Community*

Provides the conceptual guidebook and strategic road map for navigating through the perils of climate instability in the quest for economic viability and long-run food security.

— John Ikerd, Professor Emeritus of Agricultural Economics, University of Missouri

Shows that farmers, ranchers, fishers, and gatherers are humanity's best hope for connecting ourselves to nature. To help these innovators thrive, we'll need to construct regenerative food systems community by community.

— Ken Meter, author, *Building Community Food Webs*

Gifts us practical insights and plenty of examples of how we can reshape our food system from being depleting to one that is resilient and regenerative. Thank-you Laura for your practical guidance and inspiration.

—Mathis Wackernagel, Ph.D., Founder and President, Global Footprint Network, co-author *Ecological Footprint*

A brilliant, hopeful book and a call to action.

— Marianne Landzettel, journalist and author, *Regenerative Agriculture: Farming with Benefits*

Offers great hope, showing how farming based on reciprocal relationships of care—with soil, animals, ecosystems, and communities—can create sustainable farms and foodsheds. This book is timely, rigorous, practical, and wise.

— David Bollier, commons scholar/activist at Schumacher Center for a New Economics, author, *The Commoner's Catalog for Changemaking* and *Free, Fair, and Alive*

Mixing specific stories from current farmers with theory and analysis, Lengnick lays out a path for systemic, practical, and realistic transformation.

— Peter H. Lehner, managing attorney, Sustainable Food & Farming, Earthjustice, and co-author, *Farming for Our Future: The Science, Law, and Policy of Climate-Neutral Agriculture.*

Artfully guides the reader through the rationale and process that will result in more regional and resilient systems. The book shifts from the theoretical to the practical with leading practitioners across geographies and production systems that demonstrate that a more resilient agriculture is possible.

— David LeZaks, PhD, Senior Fellow, Croatan Institute

An accessible and inspiring illustration of resilience thinking and a rich collection of stories from farmers and ranchers, Lengnick helps to make the case for what is both possible and necessary in being clear eyed about and adapting to our changing climate.

— Curtis Ogden, senior associate, Interaction Institute for Social Change, lead facilitator, Food Solutions New England

Whether you are just stepping onto the resilience thinking footpath to those of us that are well versed in projected climate change impacts to our food and fiber systems, *Resilient Agriculture* offers guideposts to encourage our individual and collective journeys towards a hopeful agriculture production vision that leaves no one behind.

— Michelle Lovejoy, Landscapes Resilience, Environmental Defense Fund

When we think about how we will sustainably feed ourselves now and for future generations, the resilience of our food systems is front and center. Lengnick gives us hope by presenting the various solutions and tools applied across multiple foodways. We also get to peek into the lives of many people and communities working towards resilient food- and eco-systems across America, giving us some hope that maybe, just maybe, there is a path forward to adapt and move towards a sustainable food future.

— Jess Fanzo, Bloomberg Distinguished Professor, Johns Hopkins University

Lengnick's plea to adopt resilient thinking in how we approach the most basic of human needs—feeding ourselves—offers critical guidance in a time of unparalleled crossroads of climate, the pandemic and social justice.

— Kathleen Finlay, President, Glynwood Center for Regional Food and Farming,
Founder of Pleiades

Resilient Agriculture

Cultivating Food Systems
for a Changing Climate

EXPANDED
&UPDATED
SECOND
EDITION

LAURA LENGNICK

new society
PUBLISHERS

Cover design by Diane McIntosh.
Cover images: all author supplied except bee by Brendon Rockey,
top landscape: © Shutterstock (shutterstock_573308194).
Interior p xiii image © jihane37/Adobe Stock.

Printed in Canada. First printing May 2022.

This book is intended to be educational and informative. It is not intended to serve as a guide. The author and publisher disclaim all responsibility for any liability, loss or risk that may be associated with the application of any of the contents of this book.

Inquiries regarding requests to reprint all or part of *Resilient Agriculture* should be addressed to New Society Publishers at the address below. To order directly from the publishers, please call (250) 247-9737 or order online at www.newsociety.com.

Any other inquiries can be directed by mail to:
New Society Publishers
P.O. Box 189, Gabriola Island, BC V0R 1X0, Canada
(250) 247-9737

LIBRARY AND ARCHIVES CANADA CATALOGUING IN PUBLICATION

Title: Resilient agriculture : cultivating food systems
for a changing climate / Laura Lengnick.

Names: Lengnick, Laura, author.

Description: Expanded & updated second edition. | Includes bibliographical
references and index.

Identifiers: Canadiana (print) 20220200149 | Canadiana (ebook) 20220200181 | ISBN
9780865719507 (softcover) | ISBN 9781550927436 (PDF) | ISBN 9781771423397 (EPUB)

Subjects: LCSH: Sustainable agriculture. | LCSH: Agriculture—Environmental aspects. |
LCSH: Climate change mitigation. | LCSH: Climatic changes.

Classification: LCC S494.5.S86 L45 2022 | DDC 630—dc23

New Society Publishers' mission is to publish books that contribute in fundamental ways to building an ecologically sustainable and just society, and to do so with the least possible impact on the environment, in a manner that models this vision.

Contents

PART 4—Real World Resilience:
Stories of Land, People and Community

> **Reading Guide available for download at:**
> https://newsociety.com/pages/resilient-agriculture-2-reading-guide

Acknowledgments

Back in late 2019, my publisher got in touch to let me know it was time to think about doing a second printing of *Resilient Agriculture* or revising to create a second edition. Several conversations later, we decided a second edition was the best option. The climate science needed updating and besides, many of the farmers and ranchers featured in the first edition had continued to share their stories of climate risk and resilience with me over the years since the first edition was published. I figured the whole thing would take maybe six weeks, tops. I'll never forget the last words of advice from my publisher: "Whatever you do, don't write a new book!" Two years later, I finished what turns out to be a completely new book.

Friends, colleagues and a lot of wonderful souls well-met during the course of this work reviewed drafts and made excellent suggestions for improvement, including Mara Shea, Katy Estrada, Bob Sigmon, Mark Siler, Nancy Reed, Gina Raicovich, Bob Turner, Dave Llewellyn and Laura Anthony. Mari Stewart, Ashley Rood, Sally Goerner, Jude Wait, Kelley Tapia and Weogo Reed were especially generous with their time and thoughtful critique. Once again, Karen Gaughan never missed a beat. Many others assisted along the way by sharing their stories, challenging my ideas and offering introductions, advice and encouragement that helped me make sense of what I was learning as I worked. I thank you all.

My deepest appreciation and respect go to three fellow travelers on this journey who helped to shape this book: editor Meredith Leigh, who pushed me with confidence, courage and compassion to find my authentic voice as a storyteller and a scientist and then made sure I used it; graphic artist Caryn Hanna, who brought clarity and continuity to my ideas with her unique and engaging illustrations; and Story Medicine healer Meta Commerse, who helped me reclaim my indigeneity and find my way home.

My abiding gratitude goes out to the farmers, growers and ranchers who have so generously shared with me their experiences of growing food in these times of unprecedented challenge and opportunity. You have trusted me to tell your stories—thank you. Finally and always, to my husband Weogo Reed, thank you for everything, but especially for never, not even once, rolling your eyes when I asked if something could wait until "after the book is done."

Dandelion

— by Digger

There is
a lot to learn
in this moment
From those
who some
call weeds.
Like the settler
who names another
invader.
And
the
dandelion

Who
with the utmost
humility
Quietly defies
the hubris
of the plowman,
Repairing
with tap roots
the hardened ground
And sharing
in abundance
their nectar,

their pollen
and seeds
with the world.
Pay attention
to them,
Young Blood,
And how
they not only
survive
But thrive
through moments
of change.

PART 1

WHY THINK RESILIENCE?

I want to have hope, but I don't know.
There's also the reality of how we're treating
the planet right now. I think we are all going to be a little lost
in the dark as the weather keeps getting more and more chaotic.

— Mary Berry, Southern Exposure Seed Exchange, Mineral, Virginia

Waking Up to Climate Change

My WAKE-UP CALL on climate change was an actual phone call. It was one of those quiet moments in life that at the time do not feel too important, but looking back is a clear point of change. I remember I was grading papers on a fall afternoon, and gazing out of my office window over corn fields shimmering golden in the setting sun. The phone rang, I answered, and a woman with a snappy English accent explained that she was organizing an event hosted by the local chapter of the American Meteorological Society. "I'm calling to invite you to speak in a public lecture series on climate change impacts in western North Carolina," she said. "Sure, that sounds great," I answered. "What would you like me to talk about?" "Well," she said, "we need someone who is willing to talk about how climate change is changing the way we eat here in the mountains."

My first thought was that she had the wrong person. I'm a soil scientist, not a climate scientist. On second thought, I wondered what *was* known about climate change effects in food and farming, which was rapidly followed by my third thought: Perhaps this is a good opportunity to learn something new! And so I agreed to do the lecture. We discussed a few details, and I hung up the phone.

I will never forget that call. Thinking back on it now, I remember, in that quiet moment after the call ended that I thought to myself, I guess it's time for me to learn something about climate change.

Please understand. This was way back in 2008. It wasn't that I didn't know about climate change theory. I knew a lot about it. The Earth's energy balance. The greenhouse effect. The global processes that create weather. How fossil fuel use changes the composition of the Earth's atmosphere. Check, check, check and check—I taught these concepts every year to first-year students in my environmental studies classes. I also knew a lot about the lead antagonists in the climate change story, because I had researched the behavior of carbon and nitrogen in farming systems as a young scientist.

Besides all of that, I was one of those people who estimated my carbon footprint and made lifestyle changes to reduce it. I purchased green products, biked and carpooled when I could and voted for leaders who preached respect for our environment. I joined Transition Asheville and created an energy descent action plan for my community. I was one of the good guys.

Back then, I thought of climate change as something off in the distance—a hazy future threat. I knew it could be bad, but I also believed that we still had time to fix it. I knew it wouldn't be easy, but I felt sure we could and would begin to act on climate change before it caused too much damage.

As I prepared for that public lecture, I was shocked to learn that climate change was already happening, that farmers and ranchers across North America were already suffering increasing losses from more variable and extreme weather. I learned that cities were struggling to manage more frequent and increasingly deadly flooding and heat waves, rising temperatures were beginning to interfere with air travel and tropical diseases were moving northwards. I was surprised to learn that the recent drought I had lived through here in the Southeast was just a glimpse of a future likely to be shaped by increasing competition for water. I was astonished to learn that the Environmental Protection Agency was busy making plans for a retreat from our nation's coasts because of sea level rise. The evidence was clear: the fingerprints of climate change were already touching every part of our lives.

Learning these things changed me. I could not continue to move through a comfortable dream world that put climate change somewhere in the future. I knew that I had to shift my teaching to focus less on the processes of global warming, and more on the reality of climate change: what it is, how it is disrupting land, people, and community and, most important, what we can do about it. I needed to share what I had learned, as well as I could, and in as many places as I could:

- Climate change is happening now.
- Climate change is changing everything.

I needed to do what I could to help people understand that it was way past time for us to think about how we could adapt to climate change even as we worked to slow it down and reverse it.

In the years since giving that public lecture, I went on to lead the development of a national report on adapting U.S. agriculture to climate change, to write a book exploring climate change resilience and the future of food through the adaptation stories of some of America's best farmers, and to resign my faculty position so that I could start a small business helping people take action on climate change.

All because of a phone call one lovely fall afternoon.

Unprecedented

My first big step into the world of climate action came as an invitation to join the U.S. Department of Agriculture (USDA) leadership team responsible for producing the very first national report exploring agricultural adaptation to climate change in the U.S. As a member of the lead author team and the lead scientist on adaptation, I worked with researchers from across the U.S. to gather, review, discuss and report on the state of scientific knowledge about current and future climate change effects on U.S. agriculture as well as effective agricultural adaptation options. It was in this work that I first learned the language of resilience, the ideas and language that I now use every day, ideas like vulnerability, exposure, sensitivity, adaptive capacity, climate risk, and climate equity.

In the 18 months I spent working on the USDA report, I learned many surprising things that caused both heartache and hope in equal measure:

- Climate change is not changing weather patterns uniformly across the U.S. Where you live determines your experience of climate change.
- Climate change is not only increasing the frequency and intensity of extreme weather. Climate change is behind subtle changes in seasonal patterns of temperature and water that rarely make the news, but can be incredibly damaging to agricultural businesses.
- Climate change adaptation is not about figuring out how to adjust to a "new normal." It is about figuring out how to manage the risks created by more variable weather patterns that are likely to change at a faster pace and grow more intense through at least mid-century.

As the leadership team worked to finish the report, we spent a considerable amount of time identifying the most important lessons learned in our 18 months of work. I want to share one of these lessons here, because I think it expresses very well the unique nature of the climate change challenge to U.S. food and farming systems:

> Although agriculture has a long history of successful adaptation to climate conditions, the current pace of climate change and the intensity of the projected climate changes represent a novel and unprecedented challenge to the sustainability of U.S. agriculture.

Many concerns were raised about this statement by the authors and reviewers of the report: that it was too pessimistic; that we were getting ahead of the science; that it was not in the public's best interest to admit that we didn't have all the answers. The lead author team pushed back on these criticisms, because we felt confident that the data did support a statement of this nature. We understood it was a provocative statement. We understood that such a conclusion would cause serious concerns about our ability to sustain U.S. agriculture into the twenty-first century. And that's exactly what we hoped to do. We wanted agricultural leaders to wake up and recognize the unprecedented nature of climate change.

It has been almost a decade since we tried to alert the agricultural community and the general public about the unique challenges of climate change to food and farming. Despite the fact that the scientific evidence and our own lived experience has consistently supported our conclusion since then, we have not yet fully accepted the consequences of our failure to act on climate change.

Consider, just for one quiet moment, the word "unprecedented." It's an adjective that means never before known or experienced. Just sit with that idea for a moment: never before known or experienced…

We chose this word carefully and with a lot of intention, because we wanted to be sure that the agricultural community understood the enormity of the challenge before us. We chose the word appropriately. Never before in the entire history of U.S. agriculture have farmers and ranchers managed crops and livestock under these kinds of weather conditions. In fact, if you take the time to look at the Earth's climate history, you will see that never before in the history of agriculture—never in the 10,000 years since agriculture first began—never before have we been challenged to grow food in a changing climate. It turns out that over the thousands of years we have fed ourselves with farming, our climate has been remarkably stable.

It is difficult to grasp the reality of these times. That the weather changes we've experienced in the last decade are going to continue to grow more damaging. That the weather is not going to settle down into some new normal. It isn't easy to fully understand the fact that spring and fall weather will continue to grow more variable, that both flooding rains and drought will grow more intense and will happen more often, and that record-breaking weather will become common.

It's even harder to realize what this means for the people who feed us. How will they manage the novel risks associated with changing weather? How will they cope with the additional challenges created by these weather changes: more soil erosion, new pests and diseases, increased competition for water, more frequent and intense wildfires, more labor and market disruptions, pandemics? As I struggled to come to terms with the unprecedented nature of these challenges, I began to search for what we already knew about managing food and farming systems in highly variable conditions.

Running into Resilience

As the adaptation lead on the USDA report, it was my responsibility to collect all of the science-based information that I could find about the management of weather-related risks in agriculture. I was especially interested to learn what we knew about designing and managing agricultural systems with the ability to thrive under conditions of high uncertainty and change—including more variable weather and extremes.

As I worked to gather this information, I discovered a kind of science that explores the answers to exactly the questions that I thought were important to my work: What qualities of systems allow them to persist over time? How do we design and manage systems to thrive in changing conditions? These are the questions asked by resilience science.

Learning resilience science gave me a way to think more clearly about how agricultural and food systems respond to changing conditions. I learned a new language and new concepts that helped me to slow down, take a breath, and see farming and food with new eyes. For the first time in 20 years of work as an agricultural researcher, educator, policymaker, grower and activist, I had the benefit of a new language to help me navigate the complexity of food and farming. A new way of thinking that helped me explore with more depth the relationships between landscape, weather, community, culture and politics that shape the day-to-day decisions of farmers and ranchers. To examine more honestly how those decisions enhance or degrade the resilience of their farms, and ultimately our food system. The tools of resilience thinking helped me always keep the whole in mind even as I worked to understand each part.

Here are a few resilience thinking concepts that I have found particularly useful as I've worked to understand the resilience of agriculture and food systems:

- Vulnerability helps to identify threats that have the potential to disrupt or damage the system and to select effective responses that generate multiple benefits.

- Adaptive Strategy helps to avoid common conflicts that arise between management intention and effect that often degrade the resilience of the system.
- Response, Recovery and Transformation Capacity helps to avoid the common mistake of thinking that resilience is only about bouncing back.
- The Rules of Resilience help to keep in mind three simple rules to guide the design, assessment and management of resilient social-ecological systems.

The most valuable lesson I learned from social-ecological resilience science is that resilience is about a whole lot more than just bouncing back. We can use resilience thinking to cultivate in any kind of system—person, family, farm, city or region—the ability to respond to changing conditions, disturbances and shocks in ways that avoid or reduce potential damage. In other words, we can use resilience thinking to design and manage systems with less need to bounce back in the first place.

Resilience thinking has another gift for us, one that is especially important in this age of climate change. Any time there is a disturbance or shock that causes damage to a system, we can use the rules of resilience to identify options for bouncing forward. Every time a system sustains damage—and there is no doubt that we will experience more and more damaging events in the future—we will have to make decisions about how to repair the damage. In these moments when we will have to decide which way to go, resilience thinking can help us answer the question: "Do we bounce back, or do we bounce forward?"

This is not a trivial question. Strategies focused on rebuilding to twentieth century standards (often called "bouncing back," and lately called "building back better") suffer from the same kind of industrial thinking[1] that has created the wicked problems we are struggling to overcome today. Investments made to bounce back use up resources while leaving us stuck in the twentieth century, despite the growing evidence that twentieth century thinking is what got us into this mess in the first place.

Resilience thinking gives us a new way to think about shaping our world to thrive in change. We can use the rules of resilience to bounce forward to a way of life that recognizes that our well-being depends on the well-being of others and the planet. We can finally let go of industrial thinking and embrace resilience thinking to put us on the path to a resilient future.

Redesign for resilience involves making changes to every part of our society. It requires us to rethink some fundamental assumptions about how the world works—long held assumptions that are the foundation of modern science and technology. It requires that we wake up and confront longstanding structural inequities in our country. It means that we must let go of the toxic beliefs that fueled the industrial era: infinite growth, industrial efficiency, white supremacy, the invisible hand of the market and individualism. It means that we must redefine the meaning of success.

Resilience thinking requires a major shift in at least six major aspects of industrial thinking. I think of these six shifts as design principles that can be used to guide the transformation from our industrial present to a resilient future. These principles are expressed in terms of the current industrial thinking that must change. The list is ordered in terms of how challenging they will be to achieve from least to most challenging:

- From optimum to variable conditions—We must give up the myth that we can expect to create optimum conditions for production and consumption. While it may be comforting, this enduring myth of industrial thinking is especially dangerous given the unprecedented challenges of climate change and other global sustainability challenges.
- From efficient to redundant systems—Industrial notions of efficiency are based on the assumption that the costs of efficiency will be transferred to the public. Industrial thinking does not recognize the indirect costs to society required to support this kind of efficiency, for example public costs of soil erosion, water and air pollution, subsidized crop insurance and climate change.

- From expert to place-based knowledge—Industrial thinking does not honor local ecological, social, spiritual or cultural wisdom; rather it uses expert knowledge to overcome local conditions in a quest for uniformity. For example, the dependence of agriculture on regional water management systems in the northwest, southwest, and southern Great Plains states.
- From industrial to ecological design—Industrial thinking tends to favor simple energy and material flows over more complex flows and cycles. For example, the dependence of agriculture on simplified production settings that receive, absorb and release global flows of fossil fuel energy, fertilizer and pesticides to the environment. We must shift our thinking to focus on the design and management of systems that have the capacity to produce and recycle needed energy and materials.
- From imported to regional resources—Industrial thinking favors the movement of energy, materials, information and wastes through global and national networks. For example, society continues to rely on complex, large-scale networks to meet our needs despite growing evidence of their fragility.
- From extractive to regenerative economy—Industrial thinking ignores the real costs of industrialism's exploitation of land, people and community. For example, the full costs associated with the global industrial metabolism, such as the industrial degradation of local resources (natural, human, social, financial and built), the reliance on unpaid labor, and the value of non-market exchanges critical to community well-being, are routinely excluded from economic analyses.

A Real-World Test of Resilience

As I began to apply what I was learning about resilience to climate change adaptation in agriculture, I started to see a lot of similarities between the shifts in thinking required to cultivate a resilient food system and the principles of sustainable food and farming. The more I thought about it, the more I began to imagine how I could put the rules of resilience to a real-world test.

After finishing up the USDA report, I planned a project to gather stories from producers using sustainable practices about their experiences farming in a changing climate. Over the next two years, I made hundreds of phone calls and sent out over 400 emails to longtime sustainable farmers and ranchers growing vegetables, fruits, nuts, grains, meats and dairy products across the U.S.

Why longtime producers? Because I knew that climate change effects began to accelerate in the U.S. around the year 2000. I thought that farmers and ranchers who had been farming in the same location since at least 1990 would be the most likely to have noticed this increase in weather variability and extremes.

Why all over the U.S.? Because the changes in seasonal weather patterns associated with climate change were not the same everywhere. The Midwest and Northeast were getting wetter, while the Southwest was getting drier. Average temperatures were rising in most of the U.S., but parts of the Southeast were getting cooler. These different patterns of change meant that farm location was an important factor in how a producer experienced climate change.

Why sustainable? Because I thought that these producers offered three valuable real-world tests of resilience principles. First, environmental sustainability required that these producers use a kind of "ecological logic" in order to create a healthy farm ecosystem that produces crops and livestock with the goal of limiting the import of resources or export of wastes. Second, sustainable producers had been largely left to figure out how to create these systems on their own without the help of "the experts," so they make an interesting test of the efficacy of place-based knowledge. Finally, I knew that their choice to manage healthy farm ecosystems left these producers ineligible for most of the subsidy programs available to industrial farmers. This meant that they had no choice but to design and manage for resilience just to stay in business— they were farming without a safety net.

And so I got started. Working at night and on weekends over the winter of 2013–14, I listened as some of the best farmers and ranchers in America shared their experiences of producing crops and livestock in a changing climate. Seven years later I checked back in with these producers to hear how they were doing. I also did some additional interviews with other longtime farmers and some less experienced farmers.

This book is the cumulative result of that work. It is about what I have learned from these farmers and ranchers about real-world resilience thinking, and how I believe their stories can help us bounce forward to a sustainable and resilient food future.

Part 1: Why Think Resilience? presents current and projected changes in weather-related challenges across the U.S., explores the unprecedented nature of these challenges, and offers some climate change adaptation concepts that help to clarify options that reduce climate risk and cultivate the climate resilience of land, people and community. Read this section if you want to know more about: specific examples of unprecedented disruptions and shocks associated with climate change; how weather has changed and how it is expected to change over the next 30 years in seven different regions of the U.S.; and a useful new way to think about managing climate risk.

Part 2: The Rules of Resilience? takes a deep dive into the principles and practices of resilience thinking applied to the design and management of agriculture and food systems, which are also called agroecosystems or foodways. Read this section if you want to learn more about: the four different kinds of resilience science and which one is most useful to foodways management; social-ecological resilience thinking; some characteristics and behaviors of resilient systems; resilient agriculture design principles; the three rules of resilience; the resilience benefits of sustainable agriculture practices; and some new tools for cultivating resilience in food and farming systems.

Part 3: What Path to Resilience? explores the unique potential of resilience thinking to transform the global industrial food system. Read this section if you want to learn more about: the search for sustainable food in the U.S.; Indigenous foodways; the Good Food movement; the regional roots of resilience; and 12 things that you can do to cultivate a resilient agriculture.

Part 4: Real World Resilience explores climate change, resilience and the future of food through the adaptation stories of bio-intensive, bio-dynamic, organic, climate-smart, and regenerative farmers and ranchers growing vegetables, fruits and nuts, grains, dairy and meats throughout the U.S. Read this section if you want to hear directly from farmers and ranchers about the challenges of growing food in a changing climate, how they are managing climate risk, and how we can help them cultivate the resilience of land, people and community by changing the way we eat.

A Word About Hope

Climate change work is hard. The problem is wicked, the stakes are high, the necessary actions ask a lot of all of us, and if we fail, the future is truly horrifying. People often ask me, "How can you do this work and stay so positive?" I'll be honest with you. I find myself in some pretty dark places every now and then. Dark places that make it hard for me to continue doing this work. There are a whole host of emotions—anger, fear, regret and grief—that come along with climate change work. They can get in the way of taking action. I've seen it in my own work. I've watched it shut down the search for solutions in community—from the local to the global.

The enormity of the climate change challenge can, and often does, leave people feeling hopeless. And so I started to dig into hope, to try and understand what it is, what role it plays in our well-being—especially in times of uncertainty and change—and how hope helps us search for solutions in community. I learned that:

- Hope is a verb. It is an action word. It means "to cherish a desire with anticipation."
- Hope has been studied by psychologists for a long time. They've done so much research on hope that "hope psychology" is a recognized specialty.
- Seven kinds of hope have been identified by psychologists, including two kinds that are particularly relevant to climate change work.

The first is a kind of hope called *wishful hope*. This is the belief that somebody else—not you—will make it possible for you to achieve your "cherished desire." Somebody else—not you—will remove any barriers standing the way to your desire. That somebody else might be a higher power, or a parent, a boss, a scientist, an entrepreneur or a politician. The important thing to remember about wishful hope is that it shifts the responsibility to act onto someone or something else over which you have little or no control.

The really crazy thing about wishful hope is that it is practiced by both optimists and pessimists. Optimists don't act because they believe it's all going to be okay in the end. Pessimists don't act because they believe it doesn't matter what they do, it is definitely not going to turn out okay. Psychologists say that wishful hope leads to feelings of helplessness, despair and denial. I'd say that there's too much wishful hope going around these days.

The second kind of hope is called *grounded hope*. This is the belief that in order to realize your cherished desire, you have to work with others to achieve it. It turns out that grounded hope is very much a community-based thing. Grounded hope leads people to get together with others, to visualize a desired future, and then to work together to get there. Psychologists say that grounded hope leads to feelings of personal agency, empowerment and the acceptance of reality.

So now I have a name for the kind of hope I practice, the kind of hope that has sustained me through more than 30 years of work to explore the sustainability and resilience of land, people and community. Learning about grounded hope has helped me understand that everyone, no matter where you stand in the food system, can help put us on the path to a resilient food future.

A good place to get started is to learn about and use resilience thinking—in your family life, in your profession, and in your community. Let's begin.

2

Climate Change Is Changing the Weather

I F IT SEEMS TO YOU like weather-related disasters are happening more often, are more varied and are more damaging in more places—you're right. In fact, unprecedented disasters are happening so often these days that we have stopped finding them remarkable and often not worthy of our attention. It is increasingly clear that climate change is at least partly to blame for the steady increase in billion-dollar disasters like the lengthening wildfire season, intense drought and more frequent heavy rainfalls plaguing our nation.[1] The central and southern states are most vulnerable to these damaging weather extremes, but since 1980, every state in the nation has experienced weather-related disturbances and shocks leading to at least one billion dollar disaster.

The 2020 hurricane season set several new records: most named storms (30), most storms U.S. landfall (12), most landfall storms producing at least $1B in damages. Meanwhile, the record-setting 4.1 million acres burned by wildfires in California was more than double the previous record. The damage caused by the August 10 derecho that roared through the Midwest put the storm in the top three most expensive severe weather events in the nation—all of which have occurred since 2000.

The weather in 2021 continued to rack up another impressive list of weather firsts as record-breaking winter storms, deadly summer heat waves, widespread wildfires and extreme fire behavior, and water shortages continued to plague drought-stricken regions of the western U.S.

and northern Great Plains. In another unprecedented first, a water shortage on the Colorado River triggered a mandatory reduction in water use by the 40 million people in Arizona, California, Colorado, Nevada, New Mexico, Utah and Wyoming as well as the five-billion-dollar-a-year agricultural industry that depends on the Colorado River for its water supply.

Although extreme weather events capture the headlines because of high dollar damages to land, people and community, climate change has a more subtle side. Changing weather patterns such as warming winters, higher night-time temperatures, longer dry periods and a lengthening growing season work quietly behind the headlines to reduce yields and increase the costs and complexity of food and farming throughout our nation.

Why Think Resilience? explores what we know about climate change-related threats to agriculture in the U.S., presents a useful new way to think about managing the risks associated with these threats, and briefly introduces some of the sustainable farmers and ranchers growing food across our nation who have shared their stories of climate risk and resilience with me over the last decade.

Two Decades of Disaster

It can be easy to pass right over the latest bad weather news these days, especially when the hurricane, drought or damaging flood is in a far-off place, happening to people you don't know, in communities you will likely never visit. The thing is that the most important climate stories, the kind of stories that helped me begin to grasp just how much climate change changes everything—these stories live below the headlines. Listening to these stories has given me a new appreciation of nature's dance with seasonal patterns of temperature, light and water. A decade of living with climate change below the headlines has gifted me with a deeper understanding of how completely the people that grow our food must give themselves up to this dance—mind, body and spirit—to cultivate healthy land, people and community.

2001 Tropical storm Allison produces 30–40 inches of rainfall in coastal Texas and Louisiana and then moves slowly northeastward causing $13B in damages in June.[2]

2002 Widespread drought across thirty states from spring to fall causes $13B in damages to agriculture.

2003 Severe spring storms in the Midwest and Southeast set a new record of 400 tornadoes in one week. Hurricane Isabell disrupts fall harvests with high winds and 4–12 inch rains from North Carolina to Pennsylvania.

2004 Hurricane Ivan comes ashore in September as a category 3 hurricane to cause $29B in damages in the southeast and middle Atlantic states.

2005 A Midwest drought causes $2B in crop losses.

Richard DeWilde thinks that weather patterns began to change on his farm in southern Wisconsin around 2000, but it was the million dollars he lost in back to back 1,000-year floods in 2007 and 2008 that really got his attention. "For most of my 40 years of farming, weather used to move pretty predictably from west to east," Richard recalls. "Before the 2007 flood, I saw a weather pattern that I'd never seen before: a southerly flow bringing moisture up the Mississippi River Valley. The moisture turned in a circle before it hit the Great Lakes and then it just looped back on us and didn't stop. It just didn't move off and that's why we got 18 inches of rain. Now we have these weird looping events. I've seen it several times since and now it just scares the crap out of me when I see that loop."

2006 A record-setting July heat wave in California causes the death of 70,000 poultry and more than 25,000 dairy cows.[3]

2007 A late spring frost in the East and Midwest causes widespread damage to fruit and nut crops, while severe drought and extreme heat from June to December cause crop failure in the Southeast, Midwest and Great Plains.[4]

2008 High rainfall in June causes massive flooding throughout the Midwest, while severe drought and extreme heat reduce yields across most of the U.S.[5]

2009 A strong derecho in June damages crops across the Midwest and East. Severe drought continues in the Great Plains and Southwest.

2010 Severe storms cause high wind and hail damage across numerous states in the southern Great Plains, Midwest and Southeast in June.

Dan Shepherd cannot say he has noticed any clear trends in changing weather patterns in the 40 years he has been growing pecans near Clifton, Missouri, but the last 15 years has included several "firsts" on his farm: a total crop loss from a late frost in 2007; a levee breach in July 2008 and again in 2013; a string of three dry summers starting in 2011 and a total crop loss from hail in 2018. The combination of spring flooding followed by summer drought in 2013 created unusual stress on the pecans. "Being underwater for a week or two really set them back," said Dan. "And June 23 was the last rain we had until somewhere up in the middle of October, so that lack of water really hurt the fill on the pecans."

If you take the time to listen, farmers and ranchers in the U.S. will tell you that the sights and sounds of the seasons have started to change, that storms come in different seasons and from different directions than they did before and that the traditional weather signs and signals used for generations to guide field work are slipping out of sync with rhythms of life on the farm. If you ask them when they first started noticing a change in the weather, many will tell you that it was around about the year 2000. Climate scientists agree with them—it turns out that the pace of climate change shifted up a notch around 2000 and then did it again in 2010.[6]

2011 Drought and heat wave conditions create major impacts across the southern Great Plains. In Texas and Oklahoma, a majority of range and pastures are classified in "very poor" condition for much of the 2011 crop growing season. Hurricanes Irene and Lee, the largest and most expensive natural disasters in New York State history, cause massive flooding when they hit within one week of each other.

2012 A late spring freeze following an unusually warm spring makes 2012 the worst year ever recorded for Michigan fruit growers,[7] while 80 percent of agricultural land in the United States suffers under drought conditions that result in a record high 14 billion dollars in crop insurance payments.[8]

2013 Although drought slowly dissipates from the historic levels of 2012, moderate to extreme drought continues or expands into the western states. U.S. beef prices rise to record highs.[9]

2014 For the first time in the state's history, California growers learn that they will get no water from federal or state water projects as water supplies dwindle due to warmer winters and continuing drought.[10]

2015 The Southwest drought carries on, fallowing hundreds of thousands of acres of farmland, while an unprecedented December storm triggers historic flooding in the Midwest and sends damaging wind, snow and ice from New Mexico through the Midwest and into New England.

Jim Hayes remembers what it was like to wake the morning Hurricane Irene blasted through his valley in south-central New York in the fall of 2011. "Within a three-hour period that morning, we watched the stream that runs along the state road to our place gouge out the entire road, the whole 15 feet of macadam, just gone." He learned later that day that the flood had blown out the two bridges between his farm and a neighboring farm where he had sent his flock of 200 sheep to shelter during the storm. "With the help of neighbors, we were able to repair the bridges enough so that we were able to walk the sheep home. The damage that those storms caused was very frightening," Jim recalls, adding that the experience "really reset our thinking in a lot of ways."

Jim Koan says that frost risk has increased over the 40 years he has been growing tree fruit in Michigan. "In 2012, the whole state of Michigan had a ten percent crop of apples. Worst freeze since 1945, I believe. Then again, a year later, we had another significant freeze. Two years in a row of those extreme freezes have never been seen before in my lifetime or even by fruit growers who started growing in the thirties and forties. Spring frost is getting to be a bigger and bigger problem."

During the extraordinary 2012 drought, the Wisconsin farm where Hannah Breckbill was working went bankrupt and she was let go in the middle of the growing season. Four years later, she was renting land along the Iowa River in Iowa and her crops were destroyed by record flooding in August. "The river rose 14 feet after an 11-inch overnight rain," said Hannah. "I lost absolutely everything I was growing that year. That was rough."

2016 Drought continues in the west, while new areas of extreme drought develop in states across the Northeast and Southeast. In the fall, Hurricane Matthew brings historic flooding to North and South Carolina.

2017 Heavy precipitation during the 2016–17 rainy season leads to one of California's wettest seasons on record. Extreme drought severely damages field crops in North Dakota, South Dakota and Montana and forces ranchers to sell off livestock because of lack of feed, while across the Southeast, a severe spring freeze destroys peach, blueberry, strawberry and apple crops blooming three weeks early because of an unusually warm winter.

2018   Drought persists in the Southwest and southern Great Plains states, causing damage to field crops and forcing a sell-off of livestock in some regions because of lack of feed. Hurricane Florence hits the Carolinas in September, causing unprecedented damage from record-breaking river flooding.

2019   Record-breaking weather breaks out all over the country, starting with unprecedented flooding in the Midwest that prevents planting on a record five million acres and leaves behind flood waters that linger on farm fields all summer and into the fall. Meanwhile, in areas where crops were planted, a summer drought reduces yields and historic autumn rainfall interferes with the harvest, causing farmers to abandon tens of thousands of acres. Across the Southeast, farmers and ranchers suffer crop losses from a flash drought.

2020   Drought persists in the Southwest, a powerful derecho travels 770 miles from southeast South Dakota to Ohio in 14 hours producing widespread winds greater than 100 miles per hour, and hurricane and wildfire seasons set new records.

2021   Historic cold waves and winter storms impact many northwest, central and eastern states and set a new record of temperature departures exceeding 40 degrees F below normal from Nebraska to Texas. The drought expands and intensifies across many western states and a historic heat wave across the

Northwest shatters all-time high temperature records across the region. This combined drought and heat rapidly dries out vegetation across the west and contributes to what is likely to be another record-breaking wildfire season.

"We have several rivers that run through the ranch and during all of my childhood and younger years the rivers were always flowing," says Julia Davis Stafford, a fourth-generation owner and cattle manager at the 130,000-acre CS Ranch in northern New Mexico. "You could count on them as a source of water for livestock. That has definitely changed over the last decade. The rivers now routinely dry up in stretches. If these patterns continue to play out on these same paths it's going to be very tough in not very long."

Mark Shepard lost seventy-five percent of his chestnut crop at New Forest Farm in Wisconsin in one of the wettest years on record in 2019. "I knew this part of the country had wide swings in temperatures," Mark explains, "like a summer time high of 110 degrees and a wintertime low of 50 below zero. That's a problem, but that's not the challenge. The real challenge is to go from a spring day at 85 degrees, beautiful weather, leaves are emerging, and then have it instantly turn into freezing rain, then 18 inches of snow, and then the temperature drops to 10 below zero. That's brutal."

Keymah Durden says that more frequent and intense weather extremes have required a significant investment in backup generators at the Rid-All Green Partnership's farm in Cleveland, Ohio. "When catastrophic weather hits, we have to be able to respond very quickly because two hours without any power could mean we lose 10,000 fish that took two years to raise. That's a real bad day and we've experienced it. So we've learned to be prepared for extreme weather changes and, excuse me, it's sad to say, but we know it's not going to get any better, it's going to continue to get worse."

The Climate Change Challenge

Ten thousand years ago, in many parts of the world, our ancestors began the long walk out of the forests and grasslands and onto the farm in response to a suite of challenges not unlike those we face today. Population pressures were building, natural resources were declining and changing climatic conditions threatened their ability to feed themselves as foragers—a strategy that our species had depended on for at least two hundred thousand years. Agriculture evolved as a successful solution to these challenges as Indigenous farming cultures well-adapted to local ecological conditions slowly came to dominate traditional foodways around the globe.

At the time of first European contact, Indigenous foodways in the land that would become North America represented all three of the basic food-growing strategies found throughout the world. Horticulture evolved in forested regions (think food forests), pastoralism in grasslands (think rotational grazing) and sedentary agriculture (think irrigated grain farming) in river floodplains.[11] Viewed from a seventeenth century colonial perspective, the transformation of North American agriculture from a diverse mix of Indigenous foodways to a world leader in agricultural production was remarkably successful. Through science and technology, so the story goes, European colonists and their descendants eventually overcame every ecological and social limit to agricultural production during their 300-year conquest of the continent. It is no wonder that many in the U.S. agricultural community—farmers, growers, ranchers, researchers, extension and conservation technical advisors, policymakers and agribusiness people—are upbeat about our ability to adapt agriculture to climate change; however, their confidence may be misplaced.

Prime agricultural land is dwindling and human population continues to grow, while evidence accumulates that the way we eat threatens the stability of our planet. Amidst the abundance of the global supermarket, it is easy for us to forget about the enduring exploitation of land, people and community that makes our good life possible. And nobody seems

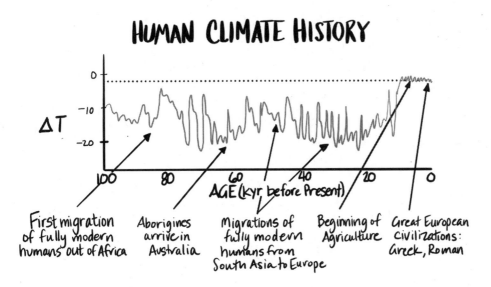

HUMAN CLIMATE HISTORY

First migration of fully modern humans out of Africa

Aborigines arrive in Australia

Migrations of fully modern humans from South Asia to Europe

Beginning of Agriculture

Great European civilizations: Greek, Roman

Figure 2.1. The last 10,000 years, the period in which humans transitioned from hunter-gathers to an agricultural society, was a remarkably stable period in the Earth's climate. This graph shows Greenland temperature variations over the last 100,000 years.[12] Credit: Caryn Hanna.

willing to remember that agriculture, from its first beginnings 10 thousand years ago to the self-driving, satellite-controlled tractor fertilizing genetically-engineered, drought-resistant corn in Iowa last spring, evolved during a period of unusual climatic stability (Figure 2.1).

As we wake up to the realities of climate change, we do not have the benefit of a stable climate. We do not have access to unlimited resources; in fact, the opposite is true: three hundred years of global industrialism has degraded every resource—natural, human, social, cultural, financial and built/technological—that has made possible previous innovation in food and farming. These resources are massively oversubscribed and most are highly degraded. Because the industrial systems that we depend on for food directly contribute to the increasing pace and intensity of climate change, we cannot burn or build our way out of this predicament. As a species dependent on agriculture for our survival, we have entered uncharted territory.

Yes, it is an unprecedented challenge, but it is not unlike the challenge that our species navigated successfully 10,000 years ago. That time, it was natural global warming that pushed us to evolve from a foraging to farming species with little awareness of where we were headed and even less awareness of what we were leaving behind. This time, we can choose instead to confront this predicament while keeping in mind all that we have learned since before the first farmers left footprints in their fields more than 5,000 years ago. We can choose to navigate this challenge with the full benefit of all that we know as a species about what it means to live well on this planet, by opening our minds and hearts to a wisdom that spans the whole of our existence—from ancient Indigenous ways of knowing to the latest twenty-first century systems science.

We can choose to let go of the industrial thinking that does not serve us well—thinking based on a view of the world that we all know is flawed. Values that we can clearly see have pushed our planet to the edge of collapse. We can choose instead to navigate the climate change challenge with a new set of values rooted in the essential truth of life on Earth: our own well-being rests in the well-being of land, people and community.

Understanding Climate Vulnerability

Growing food is a risky business. The success of every growing season hinges on a multitude of relationships between soils, crops, livestock, pests, weather, finances, regulations, markets and people. Year after year, farmers and ranchers do their best to manage these sometimes unpredictable risks inherent to their profession. In recent years, managing weather-related risks, already one of the highest risk factors in agriculture, has become even harder, as weather becomes more variable, weather extremes more common, and the number of weather-related disruptions in a single growing season increase and grow more diverse.

These changes in weather patterns have become so noticeable that they have created a new kind of agricultural risk, called climate risk. Climate risk is the additional weather-related risk created by the effects of climate change on seasonal patterns of temperature and precipitation

in a region. Climate risk is expected to become an increasingly important factor in agricultural production in the years ahead, especially if the
pace and intensity of climate change continues to accelerate as expected
through this century.

On our current path, over the next few decades, the entire United
States is expected to warm by about 2 to 4 degrees Fahrenheit (about 1 to
2 degrees Celsius). This rate of warming is substantially greater than the
rate of warming we've already experienced during the twentieth century.
There will likely be more winter and spring precipitation in the northern parts of the United States and less precipitation in the Southwest,
while summer and fall precipitation is likely to remain about the same
or decrease in most regions. Precipitation will come in fewer and more
intense rainfalls throughout the country.

Although it can be difficult to point to the exact influence of weather
on the success of an agricultural business, we can better understand and
manage climate risk by using resilience thinking to focus on the particular characteristics of agricultural operations that interact to influence
farm performance. Together these particular characteristics are a useful
way to describe the vulnerability of an agricultural business to climate
risk, as shown in Figure 2.2. By learning this language of vulnerability, the
characteristics that contribute to climate change's potential impact on
the business (exposure and sensitivity) and the business's adaptive capacity (operating context, knowledge and options, individual capability)
can be anticipated and managed to a greater or lesser degree. Thinking
about climate risk in terms of total impact and adaptive capacity helps
to clarify how management decisions made by the producer, such as production location, specific crop and livestock selections, and the design
of the production system as a whole, influence the climate risk to their
operations.

In the language of climate change vulnerability, the potential impact
of changing seasonal weather patterns on an agricultural enterprise is
defined in terms of the unique interplay between climate effects at a
specific place and time (exposures) and the response of the individual

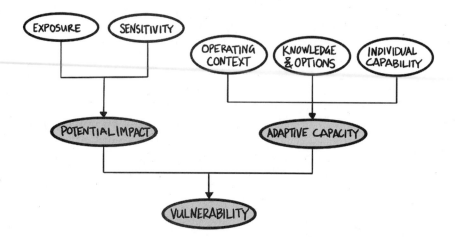

Figure 2.2. The linked factors that determine climate change vulnerability. The potential impact of climate change on a farm or ranch is determined by the interaction between its exposure and sensitivity to climate change effects. Exposure describes the type and intensity of climate change effects experienced at a specific location. Sensitivity describes the degree to which specific components of the farm or ranch system respond negatively or positively to climate-related events. Adaptive capacity describes the potential of the farm or ranch system to buffer climate change effects. Credit: Caryn Hanna.

elements of the enterprise to those effects (sensitivity). Thinking about managing climate risk in terms of exposures and sensitivities is a practical way to identify the parts of the production system at greatest risk from current and expected changes in weather.

Exposures are the specific weather-related events that occur in a particular place, such as more frequent flooding rains, dry periods or droughts, warmer winters or a longer growing season. These weather-related events can have a multitude of effects on crops and livestock that range from relatively mild—for example, higher night-time temperatures in summer that result in minor reductions in crop yield and quality—to the kind of catastrophic events that cause record-breaking damages that make national headlines.

Sensitivities are the responses of individual plant and animal species, as well as farm infrastructure like buildings, roads and fences, to weather-related effects. Infrastructure sensitivities to weather are largely determined by design tolerances that remain relatively constant. Plants and animals respond to weather events in more complex ways, but every species has an ideal range of environmental conditions that promote healthy growth and development, as well as the capacity to tolerate conditions outside of the ideal for at least a short time.

The potential impact part of the vulnerability equation is well-aligned with traditional agricultural risk management. Specific threats to the operation are anticipated, those threats with the greatest likelihood of potential impact on the operation are identified and actions are taken to reduce those risks.[13] Most of the time, contemporary agricultural risk management ends right here—at potential impact—and in doing so fails to consider what might be the most important question about the climate vulnerability of a particular farm or ranch: How does the system as a whole respond to changing climate conditions?

Thinking about managing weather-related risks through whole system response—called adaptive capacity—takes advantage of the ability of living systems to respond to challenging conditions in ways that are greater than the sum of their parts. In other words, to overcome the sensitivities of any one element in the system by drawing on living relationships within the whole that can help avoid or reduce the damages associated with weather-related disruptions, reduce the time and cost of recovery after a damaging event, or position the business to take advantage of new opportunities created by changing weather patterns. Thinking about managing climate risk in terms of adaptive capacity adds an entirely new dimension to contemporary agricultural risk management, one that can help us make the shift from managing risk to cultivating resilience.

In the language of climate vulnerability, adaptive capacity is described as the unique interplay between the social and ecological factors that limit the decisions a producer makes (operating context), the producer's access to effective climate risk management tools (knowledge and options), and the experience and management ability of the producer (individual capability).

Thinking about managing climate risk in terms of exposure, sensitivity and adaptive capacity can help farmers and ranchers anticipate and prepare for weather-related threats using the full range of risk management options available to them. For example, although no one individual can stop a drought from happening, producers can take steps to reduce the drought sensitivity of the species that they manage. They can shift to crop varieties or livestock breeds that are more tolerant of dry conditions, or adopt production practices—like dynamic crop rotation[14]—that allow more nimble responses to sudden changes in weather. Likewise, in anticipation of more frequent and intense droughts, all producers could increase the adaptive capacity of their operations by using practices that improve soil health in order to take advantage of the proven ability of healthy soils to absorb more rainfall during heavy rains and to store and release more water between rainfalls. But the benefits don't stop there. It turns out that healthy soils have many additional, well-documented benefits both on and off the farm, including reduced production costs,[15] reduced flooding downstream,[16] and drawing down carbon from the atmosphere.[17] Notice that these additional benefits promote climate resilience both on the farm and in the surrounding community that go well beyond the management of drought risk. This pattern of broad ecological, social and economic benefits to farm and community generated by on-farm investments in soil health are typical of climate risk management strategies that enhance adaptive capacity.

How has climate change affected weather patterns in the U.S.? What kinds of changes in weather are expected in the near future? How might these weather changes influence the vulnerability of agricultural businesses in the U.S.? One practical way that producers and the people that support them can explore these questions is to learn more about current and expected changes in seasonal weather patterns (exposures), how plants, animals, land and infrastructure respond to those changes, both as individuals within an agricultural production system (sensitivity) and as part of a web of relationships that determine the response of the whole (adaptive capacity). Understanding how weather patterns have changed and are likely to change in a specific location is the first step in managing climate risk in agricultural operations.

3

Understanding Exposure

THE EFFECTS OF CLIMATE CHANGE on agricultural businesses grow more damaging every year. Direct effects such as heavier rainfalls, more dry periods and drought, and more variable temperatures cause damage to crops, livestock, land and people, but the damages don't stop there. The indirect effects of changing weather patterns—for example, interruptions in fieldwork, higher pest and disease pressures, disruptions to flowering, pollination and fruiting, and increased soil erosion—combine with direct effects to create increasingly stressful production conditions for farmers and ranchers throughout the country.[1]

 Variability in precipitation has presented the greatest weather-related risk over the 20 years Jamie Ager has produced pastured beef in North Carolina, a risk that he thinks is growing. "Drought is something we've always had to deal with, but gosh, when it rains five inches, hurricane-style rains in May, it gets really difficult to operate in that kind of mud. So that's something I think has gotten more extreme. Keeping the pigs and cattle happy in long periods of soggy wet weather is probably our biggest management concern. And those are getting worse."

Rebecca Graff and Tom Ruggieri agree that extremes of temperature and moisture seem more common now than when they first started farming in northwest Missouri, but these extremes can be hidden in weather averages. "This year may have been an average year in the end," says Tom, "but what really happened was we got 12 inches of rain in July and then nothing in August and September, so if you look at the average, it looks like a normal year, but we basically seem to be fluctuating from drought to flood and back again."

"I used to say it would be one year in ten we would expect a really bad year," Colorado fruit grower Steve Ela explains, "and maybe another two or three years we would have some frost. Now I would say we have frost every year. The one-in-ten year with a ten percent crop, that still holds, but now we're having 50 percent crops many other years. When I say this was a frost-free area, it used to be that growers didn't need wind machines and other frost-protection measures and they got through just fine. Now we have the whole place covered with wind machines."

అఅఅఅ అఅఅ

Over the last century, average temperatures in the U.S. have increased, nights and winters have grown warmer, the growing season has length-ened, precipitation patterns have grown more variable and there has been an increase in the frequency and intensity of extreme weather events. These changes are not consistent across the country, but have occurred in regional patterns shaped by differences in climate, geography and land use. For example, even though producers everywhere are managing in-creased risks associated with more variable precipitation, farmers and ranchers in the Southwest are challenged by extreme drought and de-clining water supplies, even as excess moisture and flooding grow in-creasingly damaging to Northeast agriculture.

Exposure describes the kind and intensity of climate change impacts on weather patterns that have occurred or are expected to occur in a specific location. These changes could be more variable temperature and precipitation patterns, changes in seasonal weather patterns or more frequent and intense weather extremes. As climate change exposures in-crease in frequency and intensity in a specific place, so does the climate risk to agricultural operations in that region.

These changing weather patterns are being caused by a recent and rapid increase in global temperature that is strikingly different from the pattern of natural variations in climate throughout Earth's history. Global average temperature has increased by about 2°F since 1900. Climate scientists cannot find any credible evidence that this warming is the result of natural changes; instead, the evidence consistently points to human activities, especially emissions of heat-trapping gases such as carbon dioxide, nitrous oxide and methane as the dominant cause.[2]

This global temperature increase has set in motion a cascade of events challenging communities throughout the world: declines in Arctic sea ice, snow cover, alpine glaciers and lake levels, increasing sea levels, more frequent and intense drought, wild fires and heavy precipitation are just some of the weather-related changes that capture the headlines. Taken as a whole, these changes paint a compelling picture of the local consequences of global warming in the last century. These weather changes are expected to grow more intense in the years ahead, unless we find a way to reduce the level of heat-trapping gases in the atmosphere.[3]

This means that our climate is not simply shifting towards a new normal that we can adjust to and then get on with our lives at some higher, but stable, average global temperature. Because the seasonal weather patterns associated with climate change are expected to change faster and with more intensity in coming years, climate risk management will involve learning how to adapt to a constantly moving target.

Regional Changes and Expected Changes in Weather

Exposures in a specific location are determined by the interaction between regional climate conditions, local topography and natural resources, and the particular effects of climate change on seasonal weather patterns at that location. Understanding how weather patterns in your region have changed and are likely to change through mid-century can be an important climate risk management consideration. Climate change effects on seasonal weather patterns that are most likely to contribute to agricultural climate risk in seven major regions of the U.S. are described below.[4]

Northwest

The Northwest is known for clean air, abundant water, low-cost hydro-electric power, vast forests, extensive farmlands, and outdoor recreation that includes hiking, boating, fishing, hunting and skiing. The region produces a significant proportion of the U.S. supply of fruits, nuts, berries,

Table 3.1. Changing Weather Patterns in the Northwest

Observed Weather Changes
Higher temps, warmer winters, more frequent and intense heat waves, more drought and more frequent wildfires. Precipitation is more variable, especially in winter. Warming winters have increased rainfall, reduced snowpack, increased risk of flooding and soil saturation, advanced timing of spring melt and reduced summer flow in river basins fed by snowmelt. The growing season is 14 days longer.

Expected Weather Changes

Annual Temp increases. Greatest increase in the southeast, greatest summer increase in the interior, greatest winter increase in SE Idaho.	**Annual Precip** increases. Greatest increase in eastern Washington. Decrease in central Idaho and SW Oregon. Increase in most seasons; decrease in summer. Winter flooding will become more frequent.
Hot Days will increase by 10. Greatest increase in the SE.	**Dry Spells** will grow 9 to 15 days longer. Greatest increase in western Oregon. Summer drought will become more frequent.
Hot Spells will grow longer by 6 to 10 days. Greatest increase in southern Idaho.	**Wet Days** increase. Greatest increase in eastern Washington and Oregon and northern Idaho.
Cold Days will decrease by 10 to 30 days in inland regions. Little or no decrease in coastal regions.	**Freeze Days** will decrease by 30 to 40 days. Greatest decrease at high elevations. The growing season will increase by 25–35 days.

Definition of Terms: annual temp (annual average temperature), annual precip (annual average precipitation), heat and cold waves (the occurrence of 4-day periods that are hotter and colder, respectively, than the threshold for a 1-in-5-year recurrence) and extreme precipitation (the occurrence of 1-day, 1-in-5-year recurrence), hot days (annual average of days with maximum temperature exceeding 95°F), hot spells (maximum number of consecutive days with temperatures exceeding 95°F), cold days (average annual number of days with minimum temperature below 10°F), wet days (the average annual number of days with precipitation exceeding 1 inch) and dry spells (maximum number of consecutive days with less than 0.1 inch precipitation).

wheat and dry beans; leads the nation in the production of 28 agricultural products, including hazelnuts (100%), apples (75%), pears (74%), sweet cherries (73%), and potatoes (57%); and is home to important salmon and shellfish fisheries.

Changing climate conditions, especially warming temperatures and more variable seasonal weather patterns, threaten water supplies in the region as winter snowpack declines and spring snowmelt starts earlier. Fruit yields and quality are declining because earlier spring bloom creates a mismatch with pollinators and increases the risk of frost damage. Summer heat stress increases the risk of sunburn scald on apples, reduces the quality and value of berry crops, decreases forage yields and quality and negatively impacts livestock health and well-being.

The region is expected to continue to warm during all seasons. Years of abnormally low precipitation, extended drought conditions, and weather extremes will likely occur more often. The drought, water scarcity, and wildfires associated with a record-low snow pack in 2015 provide a glimpse into the future of this region if climate change continues unabated.

Southwest

The Southwest region includes some of the hottest and driest climate in the United States. Climate change has already altered factors fundamental to food production and rural livelihoods in the Southwest, particularly the shortage of water caused by droughts in California and the Colorado River Basin. Declining snowpack in the region has reduced water for people and nature, while rising temperatures intensify the impacts of the ongoing drought.

Agricultural irrigation accounts for nearly three-quarters of water use in the Southwest region, which contributes significantly to the nation's supply of vegetables (63%), fruit and nuts (69%), rice (100% short grain, 71% medium grain), milk (28%), cheese (26%) and cattle for slaughter (16%). California produces more than 80 percent of the nation's domestic supply of many popular vegetables, fruits and nuts including broccoli, carrots, celery, garlic, lettuce, tomatoes, citrus, grapes, figs, plums, rasp-berries, strawberries, almonds, pistachios and walnuts.[5] Consequently, drought and competition for water among agriculture, energy generation and municipal uses in this region pose a major risk for agriculture and food security in the U.S. Other important climate risk factors in south-west agriculture include higher temperatures, heat waves and the reduc-tion of winter chill hours.

Table 3.2. Changing Weather Patterns in the Southwest

Summary of Observed Changes

Average temps have increased in each season, most rapidly in winter. Heat waves have increased in frequency. Variability in precipitation has increased, with major droughts in the first two decades of the 21st century. Ongoing drought in the re-gion since 2000 is likely to set a new record as the longest period of severe drought since the 9th century and may signal the start of a more extreme trend toward megadrought as global warming continues.[6]

Summary of Expected Changes

Annual Temp will increase in all seasons, with greatest increase in summer. Less warming in coastal areas.	**Annual Precip** decreases. Largest de-crease in the Sierra Nevada and south-ern AZ and NM. Largest decrease in summer in parts of CA, AZ, NM.
Hot Days will increase by 5 to 20 days. Least change at high elevations. Greatest change in the south and east.	**Dry Spells** will grow longer by 15 to 25 days. Greatest increase in dry areas of NV, AZ and CA.
Hot Spells will grow longer by 8 to 16 days and by 20 days or more in the south.	**Wet Days** will increase, except for decrease in eastern CO, AZ and the Sierra Nevada.
Cold Days will decrease by 0 to 25 days. Greatest change at higher elevations in the interior north. No change in south.	**Freeze Days** will decrease 25 to 35 days. Greatest change at higher eleva-tions. Little change in coastal areas and southern CA and AZ. Growing season will increase 10 to 38 days. Least increase in CA, greatest change in the interior.

Definition of Terms: See Table 3.1.

The Southwest is expected to continue to warm in all seasons which could result in a permanent shift into a drier climate regime. Such a shift would increase the duration and severity of droughts and increase the likelihood of decadal to multi-decadal megadroughts. Precipitation is expected to decrease in all seasons and the heavy rainfalls associated with atmospheric rivers (narrow bands of highly concentrated storms that move in from the Pacific Ocean) will become more frequent and intense. Record-breaking drought, flooding and wildfires over the last decade offer a glimpse into the future of this region if climate change continues unabated.

Northern Great Plains

Agriculture has been integral to the history, economy and culture of the Northern Great Plains, home to the largest remaining tracts of native rangeland in North America, substantial areas of both dryland and irrigated cropland and pasture and mosaics of cropland, grazed grassland and forested lands. The region plays an important role in U.S. food security, producing virtually all of the nation's flaxseed, canola and sunflower seed, most of our spring wheat (70%) and dry beans (56%) and contributes significantly to domestic supplies of honey, winter wheat and beef. Nebraska leads the nation in cattle slaughter.

Agriculture in the region has benefited from a lengthening growing season and above-average precipitation over the last 10 to 20 years, particularly in the east, but these changes have been accompanied by weather variability and extremes that are increasingly outside the ranges experienced by producers in the past.

Current trends in changing weather patterns in the Northern Great Plains are expected to intensify in coming years. Warmer and generally wetter conditions are expected to increase crop production risks through more frequent disruptions to flowering, pollination and grain fill and an increase in the abundance and competitiveness of crop pests and weeds. In contrast, livestock production will likely benefit from the longer growing season but may be challenged by a decrease in forage quality.

Table 3.3. Changing Weather Patterns in the Northern Great Plains

Summary of Observed Changes

Temps have risen annually and in all seasons. Northern areas warmed at the fastest rate in the nation over the 20th century. Growing season is 6 days longer. Winter and spring are wetter and summer drier. Snowfall has decreased, particularly in the east. Drought and extreme precipitation events are more frequent.

Summary of Expected Changes

Annual Temp rises. Greatest increase in winter and summer. Greatest winter increase in NE and ND. Greatest summer increase in southwest WY.	**Annual Precip** falls in south, rising northward to a maximum in NE. Greatest increase in winter and fall. Greatest decrease in summer.
Hot Days increase by 0 to 20 days. Greatest increase in SW NE.	**Dry Spells** decrease in north up to 6 days. Increase in west up to 15 days.
Hot Spells increase by 0–12 days. Greatest increase in the south.	**Wet Days** no change.
Cold Days decrease by 10–25 days. Greatest decrease in southern MT and western WY.	**Freeze Days** decrease 15 to 21 days. Greatest decrease in the northwest. Growing season increases 20–30 days.

Definition of Terms: See Table 3.1.

Southern Great Plains

Weather in the Southern Great Plains is dramatic and consequential, putting the region's people and economies at the mercy of some of the most diverse and extreme weather hazards on the planet. Hurricanes, flooding, severe storms with large hail and tornadoes, blizzards, ice storms, relentless winds, cold waves, heat waves and droughts can result in significant loss of life and the loss of billions of dollars in property.

Agricultural production in the Southern Great Plains is extensive and diverse. The region is a world leader in the production of beef, wheat, grain sorghum, cotton and corn, and is rapidly increasing in dairy and swine production. The Ogallala Aquifer, an important source of water for agriculture in the region, is on course to become insufficient and/or completely depleted in some areas within the next 25 years. It is expected

Table 3.4. Changing Weather Patterns in the Southern Great Plains

Summary of Observed Changes

Average annual temp has increased. Hot periods are hotter, cold periods are warmer. The growing season is 6 days longer. Winter and spring are wetter, summer is drier. Snowfall has decreased, particularly in the east. Drought and extreme precipitation are more frequent.

Summary of Expected Changes

Annual Temp rises. Greatest increase in the summer and fall, least in the spring.	**Annual Precip** rises in the north and falls in the south. Little change in spring except for a decrease in TX.
Hot Days increase by 20–30 days. Greatest increase in southwest TX.	**Dry Spells** increase by 6 to 15 days in most of OK and TX. Decrease by 3 days in NE.
Hot Spells increase 8–24 days. Greatest increase in OK and northern TX.	**Wet Days** no change.
Cold Days decrease 0–10 days. Greatest decrease in the north.	**Freeze Days** decrease 0–20 days. Greatest decrease in the west. Growing season increases 15–30 days.

Definition of Terms: See Table 3.1.

that changing climate conditions, especially more frequent and intense drought and heat, will likely increasingly disrupt food, energy and water resources in the region in coming years.

Climate change is expected to increase average temperatures and periods of extreme heat will become more common. Small changes in average annual precipitation are expected; however, the frequency and intensity of both heavy precipitation and drought will likely continue to increase. The number of severe local storms, hailstorms and tornadoes may increase through mid-century. The high temperatures and increased competition for water between agriculture, cities and industry during the 2010–2015 drought offer a glimpse of the future in this region if climate change continues unabated.

Midwest

Although the Midwest is probably best known as the nation's leading producer of corn (60%) and soybeans (55%), the region also leads the nation in the production of tart cherries (77%) and pork (69%), and contributes significantly to our domestic supply of eggs (39%), cheese (37%) and milk (32%). An increasing number of higher-value fruit and vegetable crops (such as apples, grapes, cranberries, blueberries and pumpkins) also are grown in the region.

Over the past 30 years, producers in the region have been challenged by a variety of changing weather patterns. Warming temperatures have

Table 3.5. Changing Weather Patterns in the Midwest

Summary of Observed Changes

Higher annual temps, warmer winter and spring, cooler summer. Growing season is 9 days longer. Precipitation has increased, especially in spring, summer, fall. More extreme precip events during the growing season. Snowfall has decreased in the south and west, but increased in the north, in IN, and along the Great Lakes shorelines.

Summary of Expected Changes

Annual Temp increases. Greatest winter increase in northwest MN. Greatest summer increase in the south.	**Annual Precip** increases. Greatest increase in the far north. Little or no change in south. Increase in winter, spring, fall. No change to decrease in summer.
Hot Days increase 5–30 days. Change increases moving south.	**Dry Spells** increase 0–8 days. Greatest increase in north. Slight increase in south.
Hot Spells increase 5–20 days. Change increases moving south.	**Wet Days** increase. Greatest increase in states bordering Canada.
Cold Days decrease 10–25 days. Greatest change in the northwest. Least change in the south.	**Freeze Days** decrease 18–23 days. Greatest decrease in east. Growing season increase 22–30 days with greatest increase in northern MI.

Definition of Terms: See Table 3.1.

reduced crop yields, encouraged the northward expansion of new insects and diseases and increased the risk of damaging late spring frosts. Increased rainfall in spring and fall has disrupted planting and harvest, increased levels of mold, fungus and toxins in harvested grain crops, and increased soil erosion.

Over the next 30 years, expected increases in grain yields associated with increased temperatures, longer growing seasons or elevated atmospheric carbon dioxide concentrations may be offset by more frequent damage from wetter conditions in spring and fall, disruptions to crop growth and development and declines in crop quality. Extremes in precipitation are expected to intensify across all seasons, with the likelihood of both increasing heavy rain and snow events and more droughts. Increased runoff and flooding will degrade water quality in the region. Record-breaking crop losses to drought and flooding, such as those suffered by corn producers in 2012 and 2019 and to late spring frost/freeze, such as those suffered by tree fruit producers in 2002 and 2012, are likely to become more common in coming years if climate change continues unabated.

Northeast

The Northeast region is characterized by four distinct seasons and a diverse landscape that is central to the region's cultural identity and quality of life. Natural resource-dependent industries like agriculture and tourism serve as the foundation of the region's economic success. The region contributes significantly to the nation's domestic milk supply (14%). New York is a leading producer of dairy products, ranking first in yoghurt, sour cream and cottage cheese, and is among the top five states producing apples, cabbage, milk, grapes, maple syrup and cauliflower.

Agriculture in the Northeast is increasingly threatened by changing weather patterns associated with rising temperatures, more intense rainfall, increased dry periods and droughts, and declining snow and ice. Seasonal differences in temperature have decreased in recent years as winters have warmed three times faster than summers. As the proportion of winter precipitation falling as rain has increased, there are fewer days with snow on the ground, decreased snow pack and less lake ice. Prolonged periods of spring rains in recent years have delayed planting. Recent increases in rainfall intensity are the highest in the nation and have resulted in excess moisture joining drought as a leading cause of crop losses in the region.

By mid-century, winters are expected to be milder, with fewer cold extremes and frost days, less early winter snowfall, lower spring peak stream flows, and a longer transition out of winter into the growing season. Water management is expected to grow increasingly challenging for Northeast farmers as increases in winter and spring precipitation, along with more frequent summer dry periods and drought, will grow beyond

Table 3.6. Changing Weather Patterns in the Northeast

Summary of Observed Changes	
Temps have increased annually and in each season. The growing season is 9 days longer. Rainfall intensity has increased notably, particularly in the north. More intense heavy rainfall, milder winters, earlier spring melt and sea-level rise have increased the risk of flooding.	
Summary of Expected Changes	
Annual Temp increases. Greatest change moving north. Greatest change in winter and summer, least change in spring.	**Annual Precip** increases. Greatest increase in NY and DE. Greatest increase in winter. Decrease in summer.
Hot Days increase 3–27 days. Greatest change in WV and MD. Least change in north.	**Dry Spells** no change.
Hot Spells increase 1–7 days. Greatest change in WV. Least change in NY and New England.	**Wet Days** increase. Greatest change in northwest NY and northern Maine.
Cold Days decrease 6 to 24 days. Greatest change in north. Least change in south.	**Freeze Days** decrease 18–26 days. Less change along the Atlantic coast. Growing season increases 19 to 27 days.

Definition of Terms: See Table 3.1.

historical experience. These changes in precipitation amount, intensity and persistence are likely to increasingly disrupt agricultural operations—through soil compaction, delays in planting and reductions in the number of days when fields are workable, and increased pest and weed pressure—and will likely increase the social costs associated with soil erosion and agricultural runoff such as the degradation of water quality.

Southeast

The Southeast includes vast expanses of coastal and inland low-lying areas, the southern portion of the Appalachian Mountains, numerous rapidly-growing metropolitan areas and large rural expanses—all at risk from a changing climate. Agricultural production is widespread and varied throughout the region, which leads the nation in the production of peanuts (88%), poultry (66%), sweet potatoes (60%) and long-grain rice (82%) and contributes significantly to domestic supplies of watermelon (56%), citrus (45%), fresh market tomato (44%), pecans (40%) and squash (26%). While some climate change impacts, such as sea level rise and extreme downpours, are being acutely felt now, others, like increasing exposure to dangerous high temperatures, humidity and new local diseases, are expected to become more significant in the coming decades.

Since the mid-twentieth century, low temperatures are increasing three times faster than high temperatures in all seasons and the number of extreme rainfall events is increasing. Although the number of days above 95°F have decreased since the 1960s, in the last decade the number of hot nights (minimum temperature greater than 75°F) is nearly double the long-term average and the growing season has lengthened by ten days. Increasing temperatures, drought, competition for water resources and wildfire risks, together with changing conditions that favor invasive species and the northward movement of diseases, create new risks for Southeast agricultural producers.

The region is expected to continue warming at a rapid pace through mid-century. Nighttime minimum temperatures above 75°F and daytime maximum temperatures above 95°F are expected to become the summer norm. Nights above 80°F and days above 100°F, now relatively rare occurrences, are expected to become more common. The growing season is expected to lengthen by more than a month, and the number of days of freezing temperatures will decrease substantially. Precipitation is expected to decrease, particularly in summer in the south and west, but increase in the fall, particularly in the east. Water and heat stress are likely to increase crop and livestock production risks and decrease yield and quality. As the climate warms, competition for water and power is likely to increase, particularly as primary power generation moves towards natural gas and water-intensive nuclear power.

Table 3.7. Changing Weather Patterns in the Southeast

Summary of Observed Changes	
Annual and seasonal temps have steadily increased since the 1970s, particularly in summer in coastal regions. Winter temps have generally cooled over the same areas. The length of the growing season is unchanged. Precipitation has increased in fall and decreased in summer. Annual snowfall has declined.	
Summary of Expected Changes	
Annual Temp increase. Greatest change in the northwest. Least change in the southeast. Greatest increase in summer, especially in the northwest.	**Annual Precip** increase. Greatest increase in winter. Summer increase or decrease depending on area.
Hot Days increase 4–35 days. Least change in Appalachians. Greatest change in south central FL.	**Dry Spells** increase 0–12 days along the Gulf Coast. No change elsewhere.
Hot Spells increase 4 to 20 days. Least change in Appalachians. Greatest change in the west.	**Wet Days** increase. Greatest increase in Appalachians.
Cold Days decline to zero.	**Freeze Days** decrease 0–25 days. No change in southern FL. Change increases moving north. Growing season increases 0–30 days. Least change in southern FL. Greatest change in north and southern LA and AL.

Definition of Terms: See Table 3.1.

4

Understanding Sensitivity

A S WEATHER PATTERNS increasingly vary, understanding and taking advantage of what we know about how environmental conditions influence crop and livestock growth and development is an important step towards managing climate risk. *Sensitivity* describes the degree to which the individual elements of the production system—the crops, livestock, pests and diseases, land, infrastructure and people—are affected by weather-related exposures.

The sensitivity of agricultural infrastructure—buildings, roads, fences, levees, machinery and the like—is determined by the tolerances included in its design. For example, a building is typically designed to withstand a certain snow load or wind intensity, and vehicles have a maximum load. The levees on Dan Shepherd's farm in Missouri were overtopped by flood waters for the very first time in 2013—nearly 40 years after they were constructed. The flood levels on the Chariton River that year may have risen above the maximum flood level the levee was designed to hold back.

The sensitivity of the living elements of an agricultural production system are more complex because the ability of plants and animals to adapt to changing conditions varies. This variation can depend on the given species, cultivar or breed, and age. Every species of plant and animal has an ideal range in environmental conditions (temperature, moisture, light, nutrients, etc.) for normal growth and development and a broader range in these conditions that can be tolerated for short periods without permanent damage or death. This sensitivity to environmental conditions often changes over the lifetime of an individual plant or animal.

Kole Tonnemaker remembers that his first decade or so managing fruit production at Tonnemaker Hill Farm in eastern Washington went pretty smoothly weatherwise, although major crop losses in 1985 and then again in 1991 and 1992 from extreme weather got his attention. "We grow stone fruits, which are more sensitive to variable weather in the winter and spring. They bloom earlier and they're more susceptible to winter kill, so that's always something that's on our mind. Once the stone fruits—cherries, peaches and nectarines—have broken dormancy and started to lose their cold hardiness, they cannot reacquire it. They're very vulnerable to temperature variability. If you get a warm week in January and they start to lose dormancy, and then all of a sudden you get a cold spell in February, damage to the fruit bud is a big concern. The apples and pears, which are pome fruits, are less sensitive because they can reacquire cold hardiness if temperatures fall again after a short warm period in winter or spring."

∽◌◌∽ ◌◌◌

Is any part of the farm or ranch in a flood plain or open to prevailing winds? What kind and ages of livestock are managed in the operation? Is the farm subject to existing non-weather-related stresses such as poor quality soils, inadequate water resources, unmet labor needs or difficulty securing credit? What are the chill hour requirements of perennial crops? Are current exposures changing resource demands by requiring more inputs, new equipment or a change in production practices? How is the operation likely to respond to expected changes in exposures? Asking these kinds of questions about the response of individual elements within the whole farm to changing weather patterns is a useful starting point for practical adaptation planning because quite a lot is known about resource conditions required to support the sustainable production of crops and livestock.[1]

Crops and Livestock

In terms of climate change vulnerability, plant and animal sensitivities are the characteristic responses to weather-related exposures of a group of closely related species or an individual species. Plant cultivars or livestock breeds of the same species can also demonstrate characteristic sensitivities that are often different from one another. For example, through plant selection and breeding, producers have the benefit of new cultivars of vegetable, fruit, grain and forage crops that tolerate environmental stresses outside the typical range for that species. As pasture-based livestock production grows in popularity, producers are rediscovering heritage livestock breeds that thrive in more variable environmental conditions.

When Gabe Brown shifted to intensive grazing management at Brown's Ranch in south central North Dakota, he saw a big improvement in the ability of his cattle to adjust to weather extremes. "The way we manage our livestock operations, there are very few weather-related events that will affect our animals. We used to calve in February and March, so shelter, animal health and feeding during the cold were all a problem. Now, we calve in late May and June out on grass and that is a healthy environment for them. Due to our selection process the cattle are now more adapted to our environment. We raise cattle in a much more natural way now. The environmental extremes do not affect our livestock as much anymore."

As climate change disruptions grow more frequent, Mark Shepard's nursery business at New Forest Farm in southwest Wisconsin has grown exponentially. "There are lots and lots of growers that are interested in 'reality-adapted' cultivars," Mark says, "whether they are organic or chemical. Our conventional horticultural crops are all genetic weaklings. They're bred to be raised in a controlled environment. The controlled environment is not reality and it costs a lot to create it. It costs labor and it costs infrastructure. It costs time and it costs money.

It costs inputs. That's a cost that we don't need if we breed and select our cultivars to actually survive in reality. I look for the varieties and cultivars which survive with sheer total utter neglect."

꒰ꑰ꒱ ꒰ꑰ꒱

Already, heat waves and hot night-time temperatures are disrupting fruit set in vegetable crops and reducing yields in grain crops in some parts of the country. Warming winters and more variable spring weather are also disrupting crop development, especially in crops like winter grains and perennial fruits and nuts that require a dormancy period before flowering. And extremes of precipitation—both flooding rains and drought—are creating challenging growing conditions for crop production everywhere.

At Peregrine Farm in central North Carolina, increasing summer temperatures and drought seemed to be causing a decline in the productivity of some crops, according to Alex Hitt. "We've had some heat-related pollination problems in tomatoes, squash, beans and cucumbers. Temperatures are just too hot for fruit set." High fall temperatures have also caused problems, sometimes scalding the tomatoes on the vine. Drought has interfered with normal plant development too, causing time to maturity to become more irregular.

In 2012, temperatures at Fuller Farms in east central Kansas were so much warmer all through the spring and summer that everything was about 30 days early, according to Gail Fuller. The winter grains were stressed during the grain filling period by the hot, summer-like conditions in spring, while corn and soybeans were stressed throughout the summer by excessive heat and high humidity. "It was just over 100 degrees every day, day after day that summer," Gail recalls. "If we could have cooled off at night and let those plants relax a little bit, we probably would have had a better chance. It's really the night-time temperatures that got us more than anything. Obviously 110 degrees in the day is not anything to like, but when they can't cool off at night it makes it so much tougher the next day."

Ira Wallace says that more variable weather has required increased attention to the timing of harvest and processing of seed crops at the Acorn Community's farm in Virginia. "Knowing when the seed is close enough to maturity to harvest, even when they haven't fully dried up, is more important now. We've learned we can't leave seed crops to dry down in the field like we used to, because we never know when we might get a heavy rain. So we harvest seed as soon as we can and do the final drying indoors."

⌒◯◯⌒ ◯◯◯

Plants are most sensitive to extremes of temperature and moisture during the reproductive phase, when flowering, pollination, fertilization and fruiting take place. Animals are most sensitive to extremes during pregnancy and just after birth. The damaging effects of temperature extremes are increased by moisture extremes, which often occur together, such as a heat wave accompanied by drought. Seasonal weather patterns and extremes play an important role in determining crop and livestock growth rate, yield and product quality. Generally speaking, temperatures between 40 and 80°F and 65 to 90°F are optimal for the growth and development of cool-season and warm-season grains and vegetable crops, respectively.

Temperature sensitivity in livestock is determined by a variety of factors including species, breed, cultivar, hair-coat thickness, body condition, wind, humidity/moisture, shelter, diet, access to water and period of acclimation. Generally speaking, temperatures within a range of between 40 and 75°F are optimal for livestock growth and development. Animals can acclimate to gradually changing temperatures, but long periods of temperature extremes, or extreme fluctuations in temperature, will reduce productivity and can sometimes result in death. Combinations of wet conditions, cold temperatures and wind during the spring birthing period, a particularly sensitive time in the life of an animal, can present significant challenges to producers managing livestock outdoors.

 Iowa pastured-beef producer Ron Rosmann suspects that increased weather variability, especially extreme temperature swings in winter, are creating stress in his cattle. "It used to be that the cattle could tank up on feed because you could count on the cold spell hanging around a couple weeks," Rosmann said. "Now we've got 50-degree, 60-degree swings in a matter of days, plus you throw the wind in. That's the biggest change I've seen in my lifetime. We've always had those variations but not continually. You'd have longer periods in between. These extreme swings are hard."

Soil and Water

Soil and water resources are extremely sensitive to changing seasonal weather patterns and more frequent and intense weather extremes. Wind and water erosion degrade soil quality. As more variable precipitation narrows the window for fieldwork such as planting, cultivating and harvesting, the risk of soil degradation increases because producers may be forced to carry out operations when the soil is too wet or too dry. Longer growing seasons and warmer winters, combined with more heavy rainfalls also increases the risk of soil and/or nutrient loss to surface and ground waters, particularly during periods of heavy rainfall.

More rainfall in the spring and drier conditions in late summer and fall over the last decade have complicated crop management on Russ Zenner's dryland grain operation in Idaho. During spring 2011's record-breaking wet, Zenner created some compaction problems by planting on soils that were extremely wet, and he has been struggling to restore those soils ever since. He believes that the damage done in 2011 has increased soil-borne diseases on his farm.

"Longer term, no-till is not providing the results that we initially thought we might see," Russ explains. "We're still trying to identify what's holding us back. Part of it is disease. The other part of it may be the amount of manmade chemistry we're using in these systems. Many of us doing long-term no-till have wondered about this. I've questioned glyphosate, even though it is an integral tool in our no-till systems. We've got soils now that had twenty to twenty-five years of repeated glyphosate applications. There's no work being done on long-term implications of glyphosate on soil biology, and how it may possibly impact root diseases, so we just don't know. There are a lot of interactions of this manmade chemistry in the soil that we just don't understand."

Thinking about the future, Bob Quinn is concerned about how more variable precipitation might affect crop production at Quinn Farm and Ranch in north central Montana. "Water is always a challenge because we're in a semi-arid region. That means we're always short. Normally that's the limiting factor in crop production here, so we're always looking for ways to conserve water, to catch more water that falls in the land." Bob is always thinking about new ways to conserve water because he believes that conflicts over water supplies are likely to grow in coming years. "I think that water is going to be the next big battleground. It'll make the fuel crisis look like Disneyland. If we can figure out how to grow at least some basic food crops, grains, vegetables and fruits, without a lot of water, that will be a huge benefit, because we may be forced into that at some point."

ﻌﻌﻌﻌ ﻌﻌﻌ

Water availability is widely recognized as the most critical near-term resource sensitivity for agriculture. Increased variability and more frequent and intense extremes of temperature and precipitation are already creating significant disruptions in water supplies throughout the country. This challenge is expected to grow more pronounced in coming years. Rising temperatures, reductions in snowpack and shifting precipitation patterns have begun to alter the patterns of agricultural demand for water as well as its availability and cost throughout the United States.

When it comes to water management, climate is not the only source of pressure on producers. In many parts of the country, they face competing societal demands for water as urban and residential populations grow and land-use patterns change. Some of the largest demand increases are expected in regions where groundwater is the main source of water, such as the Great Plains and parts of the Southwest and Southeast. Competition for water among agricultural, industrial and public uses has reduced water supplies to agriculture in the Southwest, and a similar situation is on the horizon in the relatively water-abundant Southeast.[2]

 "We have been in and out of a drought since 1998, more in a drought than out, so water out here is everything," says Jacquie Monroe about the challenges of growing vegetables in the Front Range of Colorado at Monroe Organic Farms. "We have to irrigate in order to get a crop." Her son is the fourth generation working on the farm, but competition for water in the region makes it difficult to imagine a lifetime in farming. "We are very concerned in the future about our water rights and whether or not we're going to be able to get our water," Jacquie explains. "The cities are buying the water off the farms and taking it back to the city. They say that 700,000 acres is supposed to be dried up in the next 10 or 15 years. That water will never go back to those farms. Once it's gone, it's gone forever."

Weeds, Insects and Disease

Thinking about agricultural sensitivity to weather-related exposures grows more complicated when the effects of changing temperature and precipitation patterns on crop and livestock pests and diseases are added to the story. Generally, changing seasonal weather patterns to date have increased pest populations and disease incidence and reduced the effectiveness of cultural and chemical pest and disease management practices.

Complicating matters is that an increased use of pesticides creates a more favorable environment for swift pest and disease reproduction—

a well-documented situation known as the pesticide treadmill[3]—and so is expected to speed up the development of weed, insect and disease resistance to chemical controls.

With warming temperatures, current weed distributions will likely shift north, changing the mix of weed species as some common weeds leave the farm or ranch and new weeds arrive. Fluctuating water availability will have highly variable effects on weed success, because drought tolerance varies widely among both weed and crop species. As a general rule, farmers and ranchers can look to comparable production systems to the south or in regions where current climate conditions are similar to those expected in their region for insight about troublesome weed challenges ahead. Because of the complexity of climate change impacts on weed behavior, monitoring weed populations and the efficacy of weed management efforts will become more important under changing weather conditions.

Many of the observations about climate change effects on weeds hold true for insect pests. Warming temperatures and longer growing seasons can lead to more pest management challenges in crops and livestock because pest species can produce more generations in each growing season. These conditions also change the mix and relative numbers of existing pest populations and create opportunities for the introduction of novel species. For many pests, rising temperatures have increased winter survival without reducing populations in higher summer temperatures. This has led to higher pest populations and a northward expansion of range without a southern retreat.

 Jacquie Monroe says that another change that has come with the long drought at Monroe Organic Farms has been more challenging weed management. "Weeds are starting to go crazy out here. We're finding some of them are becoming very invasive. I can give you two examples. We've always had what's called goatheads. It's a small weed that grows very low to the ground and has a burr that sticks in your tires and in your shoes. It used to be only in certain parts of the farm, but now it seems to be going everywhere. The other one is sunflowers. We've never had sunflowers here before the drought. They are literally taking over all of our ditches. Anywhere that you can't mow or get to, they grow like trees, and we can't get rid of them. These weeds are getting to be a problem for us."

Thinking back over his 40 years of fruit growing at The Happy Berry in South Carolina, Walker Miller identifies managing pests and diseases as his most difficult production challenge, partly because it is constantly changing. "Weeds, insects and diseases—the list is quite long and invasive species are growing more challenging in this area as the climate has changed. The kudzu stink bug, the spotted wing drosophila and the brown marmorated stink bug are here. The spotted lantern fly and others are headed our way. Robins are also a perennial pest for us. They arrive in late July and leave in late August. In 2020 we were visited for the first time by blackbird flocks. I don't know what that portends."

∽◌◌∽ ◌◌◌

Because of these uncertainties, and a lack of research-based information, regular on-farm monitoring of insect populations will become increasingly important as a way to keep track of changing pest populations and the performance of insect-pollinated crops. As with weed management, to get a glimpse of future insect pest challenges producers can look to comparable production systems under climate conditions that are similar to those expected for their region.

Like weed and insect management, disease management is likely to become more difficult and costly as climate risk intensifies. Changes in seasonal weather patterns, more extreme weather events and increasing atmospheric CO_2 levels are likely to cause changes in the timing, spread and ability of disease organisms to cause infection. This will complicate cultural and biological management, and if producers increase their use of chemical control options as a result, resistance to chemical controls will emerge more rapidly as the climate continues to warm. The introduction of new crops and livestock that can better deal with climate risk may have the unintended consequence of increasing the risk of introducing new diseases, and may create new opportunities for existing diseases as well.

 According to central North Carolina vegetable grower Ken Dawson, downy mildew, a devastating disease of melons and cucurbits, is making the trip up from Florida to Maple Spring Gardens earlier than in years past. "We could safely grow susceptible crops like cantaloupe and winter squash up until sometime in August, and then those diseases would come. In the last decade or so, downy mildew has started appearing in North Carolina in June. We had to shift our plantings of susceptible crops earlier by at least a month because if we plant it later, it all dies before it matures."

Jim Hayes has noticed that parasite pressures in his sheep flock have increased as winters have gotten warmer and wetter at Sap Bush Hollow Farm in southern New York. "We've been here a long time and the winters are not anywhere near as severe as they used to be. About 15 years ago or so, we started really having problems with heavy parasite loads."

❧❦❧❦❧

In addition to the impact on crop health, regional warming and variability in seasonal rainfall patterns may also change the spatial and temporal distribution of livestock diseases sensitive to temperature and moisture, such as anthrax, blackleg and hemorrhagic septicemia, as well as increase the incidence of ketosis, mastitis and lameness in dairy cattle. This may be compounded further by the fact that weather-related stress can compromise livestock resistance to disease.

People, Community, Money

It is increasingly clear that climate risk will increase operating, maintenance and overhead costs in many agricultural businesses because of increased production and marketing costs, disruptions to needed production inputs, services and labor, and reduced product yields and quality. New challenges associated with changing weather patterns will likely add to management costs and increase the stress of managing agricultural operations.

 Western Colorado fruit grower Steve Ela is concerned about the nature of the climate risks facing Ela Family Farms. "We're investing a lot of money into planting new trees. It costs somewhere around eight to twelve thousand dollars in the first year to plant a new acre of trees and it's a ten- or twelve-year payback period if we do everything right. So any time you put more risk in that equation, it's scary. You can't really stop planting out of fear, because if you don't renovate, plant and keep moving forward, pretty soon you're going to have a bunch of old trees, with nothing coming up beyond them to support the farm. It's a Catch-22 and that is unnerving, it worries me. It's certainly something I've thought about quite a bit."

Regenerative livestock producer Jamie Ager views dealing with weather variability at Hickory Nut Gap in western North Carolina as just a normal part of farming. "I used to get more worried about it. There are so many variabilities in farming that you can get all stressed out. Part of being a successful farmer is probably just your head space as it relates to these things. But the fact that we're having more unpredictable weather creates a low level of constant worry that can be taxing on the spirit."

࿔࿔࿔ ࿔࿔࿔

Thinking through these stories and others shared with me about how changing weather patterns can disrupt every part of an agricultural operation helped me begin to see the unique nature of the risks that farmers and ranchers manage and how our failure to act on climate change has increased these risks, not only for farmers and ranchers, but for anyone who eats.

Understanding Adaptive Capacity

EVERY AGRICULTURAL OPERATION is located in a specific place with a unique set of ecological and social conditions, all of which influence the choices made by producers about the design and management of their business. For example, ecological conditions such as local topography, climate, seasonal weather patterns and the quality of local natural resources—particularly soil and water—shape the options that producers have about the kinds of crops and livestock to produce and how to produce them.

Similarly, social conditions both on and off the farm or ranch—such as family relationships, personal and business goals, government regulation and support programs, access to financial and technical resources, marketing opportunities and community support for agricultural businesses—play an important role in management decisions. Within the limits created by this mix of ecological and social conditions, producers can choose to manage relationships between individual elements of their operation in a way that improves the performance of the whole—a quality that is known as the *adaptive capacity* of the production system.

 Colorado fruit grower Steve Ela began direct marketing about 20 years ago to retain more control over the pricing of his products at Ela Family Farms. In a region that has lost 75 percent of its fruit growers over the last 25 years, he said that direct marketing, which provides high returns and product flexibility, has contributed to his continued success. "With the direct marketing more control on price means we don't have to hit a home run every year to still be viable." Direct marketing also provides his operation an additional risk management benefit because it allows him to plant varieties that were not valued by the wholesale markets that he used to sell to, but have qualities that reduce climate risk to his operation and also increase the value of his fruit in direct markets.

୭◌◌୭ ◌◌◌

Adaptive capacity describes the ability of an agricultural operation as a whole to cope with the consequences associated with changing environmental and social conditions, and also to take advantage of new opportunities they may present. Managing a farm or ranch in this way—as a whole network of relationships that are more or less responsive to changing ecological and social conditions—is admittedly complex. It requires producers to think not only about the sensitivities of the individual components of their operation, but also to think through how the relationships between the components influence the ability of the whole operation to respond to disturbances or shocks in ways that avoid or reduce damage. In terms of climate change vulnerability, as adaptive capacity increases, climate vulnerability falls and climate risk is reduced. Like vulnerability, adaptive capacity is determined by a mix of conditions created by actions taken both inside and outside an agricultural operation.

Enhancing the adaptative capacity of a farm or ranch requires making strategic management decisions that keep in mind all the moving parts of an agricultural operation. Some of these parts are simple and stable no matter the weather—for example, buildings and machinery. Other parts are more complex, but still predictable, like the growth and development pattern of a specific crop cultivar or livestock species. Some relationships on the farm are more complex—for example, the ecological conditions governing the storage and release of plant nutrients in the soil or social conditions that determine the costs of production and product value. Many others are unknown or unknowable. Making the shift from thinking about specific sensitivities to thinking about how all the different parts of the operation interact to support the whole is made a lot easier by using management strategies that take advantage of the fact that farms and ranches share many characteristics with natural ecosystems.

Farms and Ranches Are Ecosystems

Fundamental to life on Earth, ecosystems are the biological foundation for the well-being of land, people and community. An ecosystem is a community of living organisms interacting with each other and the nonliving environment through the flow of solar energy and the cycling of the basic building blocks of life like carbon, nitrogen and water to live and produce the next generation.

Ecosystems can be described in terms of some key structural and functional properties that largely determine their health and productivity. The structural properties—such as species diversity, vegetative architecture and the food web—describe the physical relationships between the organisms that inhabit the ecosystem. Functional properties describe the dynamic processes that capture, move and store energy and materials in the ecosystem, regulate the populations of organisms that inhabit it and shape the development of the system over time. These are the fundamental biological processes that support life on Earth: photosynthesis, herbivory, predation and decomposition.

These ecosystem processes are often referred to as "ecosystem services," to highlight the many environmental and social benefits that are routinely produced by healthy natural landscapes. Regional ecosystems provide a suite of natural resource services to the farms and ranches that reside within them. The regulation of soil quality, water quality and quantity, waste processing, the suppression of pests, pollination and climate protection (e.g., flood control, wind and temperature moderation) are just a few of the ecosystem services that directly influence the adaptive capacity of agricultural operations.[1]

Agricultural operations can be thought of as a special kind of ecosystem, called an agroecosystem, to recognize the high degree of human influence on ecological structure and function. Just like in a natural ecosystem, energy flow and material cycling can be explored between components of a field, a whole farm or ranch or even an entire agricultural region. At each scale, the physical components of the agroecosystem can be described and measured, the relationships between different components investigated and emergent properties[2] like soil, crop and livestock health, profitability, sustainability and resilience explored.

"A testimony to the resilience of our farm system is the fact that we have never lost a crop to pests in 30 years now of no pesticides. We have never even come close to having insects or disease or anything destroy our yields," says Ron Rosmann about his integrated organic crop/livestock farm in Iowa. "We've had stable, very good yields during all that time. That's always something that people tend to look at you with, 'Huh? I find that hard to believe.' People don't believe it. Of course, I'll be the first to say we have a lot of things we could be doing better, but still, overall, the resilience shows itself in our system. Our soil quality shines here. It really does. And we work to enhance those ecosystem services, as they're called, by continually planting more trees, more shrubs, more crops for pollinators, more windbreaks and more wildlife habitat. The diversity is what will continue to play a big role for us."

Gabe Brown says the most effective climate-risk management tool he has is the capacity of the healthy soils at his ranch in South Dakota to buffer more variable rainfall and temperatures. "After no-till for twenty-plus years, very diverse crop rotations, cover crops, plus livestock integration, we've improved the health of our soil to the point that the infiltration rate, the water-holding capacity and the nutrient cycle are totally different now. I can easily go through a two-year drought and it does not affect our operation to any great extent because the soil is so much more resilient. Now you're still going to have some swings in yields with annual precipitation, but it does not affect crop yields to the extent that it used to. If you have a healthy resilient soil resource and a functioning water cycle then your crops and livestock are not nearly as susceptible to these extremes."

Managing Agroecosystems

Sustainable farmers and ranchers are ecological engineers—they design and manage a constantly changing seasonal interplay of relationships among species to produce marketable goods and community services. To comprehend and manage this complexity, sustainable producers focus on managing four fundamental ecological processes that sustain the health of all ecosystems, including agroecosystems.[3]

- Energy flow is managed with practices that maximize the capture—in space and time—of solar energy through photosynthesis and efficiently cycling this energy through the agroecosystem.
- Water cycling is managed with practices that maximize the capture, storage, use and release to groundwater of all the precipitation that falls on the agroecosystem.
- Mineral cycling is managed with practices that maximize the on-farm production and recycling of crop nutrients through the agroecosystem.
- Community dynamics (the relationships within and among different species in the agroecosystem) are managed with practices that generate the biological diversity needed to maximize the on-farm production of ecosystem services that promote energy flow, water and mineral cycling, as well as many other services such as pollination, pest suppression and climate protection.

These four processes, each essential to the production of agroecosystem services, are linked to each other in complex ways. A change to the functioning of one process will likely change the functioning of the other three, and the healthy functioning of each process is required to support the healthy function of all the others.

Take the production of nitrogen, an essential plant nutrient, as an example. In order to supply nitrogen in the amounts needed by crops through agroecosystem services, a producer must cultivate sufficient community dynamics to capture nitrogen from the atmosphere, store it in plant tissues, make it available for crop use and then recycle it. Each of these processes requires the activities of diverse groups of organisms that include plants, animals, bacteria and fungi. All organisms in the agroecosystem require a sufficient flow of energy, water and nutrients to capture, store and release nitrogen. All four processes work together to support ecosystem health, which is the foundation of agroecological sustainability.[4]

These four ecosystem processes are easy to observe in agricultural production systems and can be managed quite effectively with currently available knowledge and technologies. The regular observation of these four ecosystem processes provides sustainable producers with a useful way to monitor the effects of management decisions on the adaptive capacity of the agroecosystem.

Cultivating Adaptive Capacity

The adaptive capacity of an agricultural business is determined by three characteristics that together describe the likelihood of successful performance under conditions of uncertainty and change. The *operating context* describes the ecological and social resources both on and off the farm that together shape the choices available to farm management. The *individual capability to act* describes the beliefs, values and attitudes of the management team and their ability to manage changing conditions. *Existing knowledge and options* refer to the state of knowledge—generated both on-farm and in society—about how to manage change. Because each of these characteristics that play a role in adaptive capacity are shaped by actions taken both on and off the farm, each is place-based and production system specific.

The Operating Context

Creating a picture of the operating context of a farm or ranch involves taking a look at the many social and ecological conditions that create barriers and opportunities for producers to design and manage a production system for high adaptive capacity (see Figure 5.1).

Because agricultural businesses are nature-based, their performance is dependent on the quality and availability of natural resource assets, particularly soil, water and a stable climate. This relatively straightforward

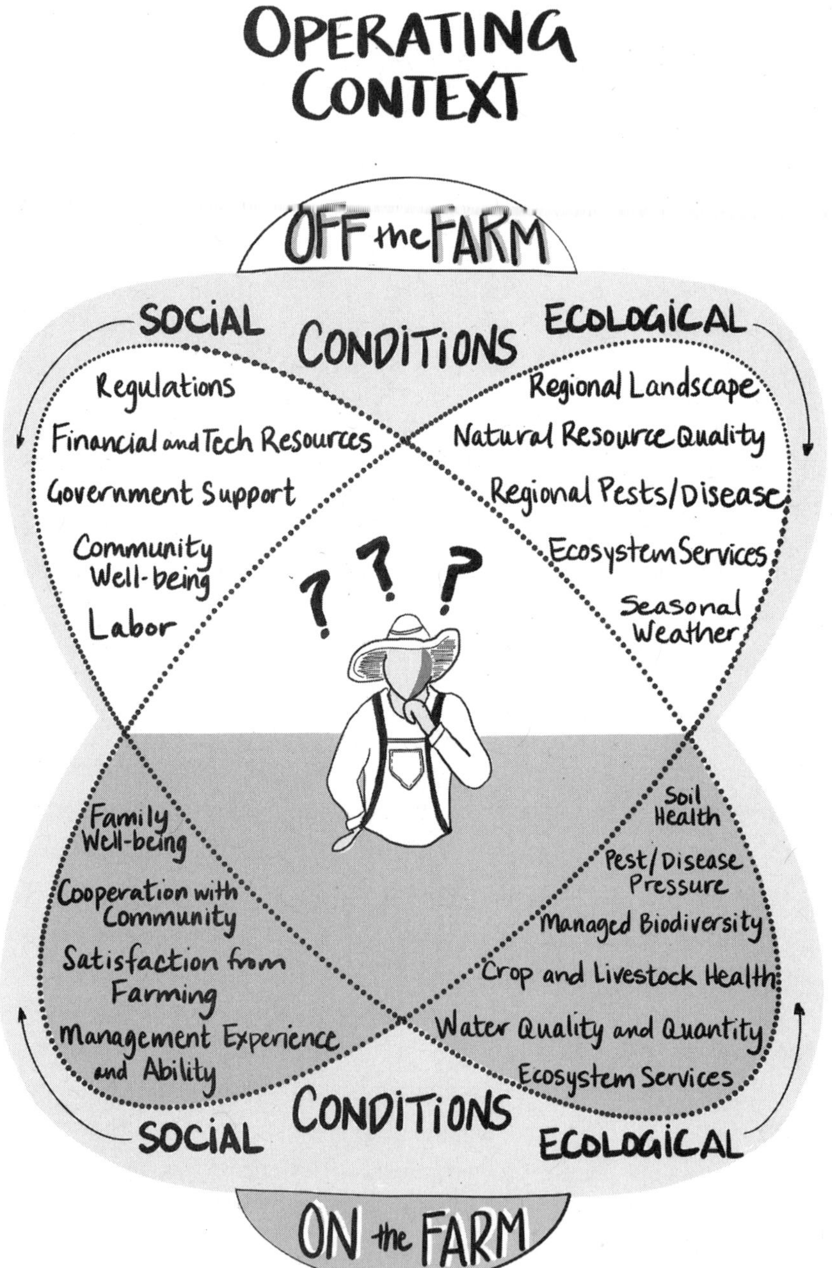

OPERATING CONTEXT

OFF the FARM

SOCIAL CONDITIONS **ECOLOGICAL**

Regulations

Financial and Tech Resources

Government Support

Community Well-being

Labor

Regional Landscape

Natural Resource Quality

Regional Pests/Disease

Ecosystem Services

Seasonal Weather

? ? ?

Family Well-being

Cooperation with Community

Satisfaction from Farming

Management Experience and Ability

Soil Health

Pest/Disease Pressure

Managed Biodiversity

Crop and Livestock Health

Water Quality and Quantity

Ecosystem Services

SOCIAL CONDITIONS **ECOLOGICAL**

ON the FARM

Figure 5.1. There are many social and ecological conditions that interact on every farm or ranch to create barriers and opportunities for agricultural producers. The choices each farmer or rancher makes about how to design and manage their operation influences the adaptive capacity of their operation. Credit: Caryn Hanna.

relationship between agricultural performance and natural resource conditions is complicated by the many social interventions in the U.S. food system that promote (for example) the use of industrial production practices, the exploitation of farm and food system laborers and the overconsumption of highly processed, nutrient-poor foods, and so help to shape the operating context of U.S. farms and ranches.[5] Ecosystem services are increasingly compromised by social activities that degrade ecosystem health, such as the intensive use of agricultural chemicals and fertilizers, overgrazing, excessive tillage, overdraft of ground and surface waters, improper waste stream management and the burning of fossil-fuels.

Agricultural water use is a good example of the profound effect of social intervention in natural resource systems. In much of the western part of the United States, natural rainfall is too low for the production of many cultivated crops and improved pastures. Over the last century, agricultural production systems dependent on irrigation evolved in this region with the support of favorable agricultural policies that subsidized the development of irrigation infrastructure and the delivery of water supplies. Unsustainable water use by agriculture and industry, unsustainable development policies and rapid population growth in the region's metropolitan areas have degraded soil and water resources and reduced the supplies of water available to agricultural production. Continued access to sufficient quantities of high-quality water is a key ecological condition that shapes the operating context of agriculture in the southwestern United States.[6]

Current social responses to the damaging effects of climate change on agricultural operations are likely to amplify these and other existing ecosystem threats—for example, increased use of synthetic fertilizers and herbicides, increased demand for irrigation water, increased use of energy and water for cooling protected growing facilities and increased use of pesticides.[7] The cost and availability of basic agricultural inputs—land, energy, water, seed, fertilizer and labor—as well as access to profitable markets and changing consumer demands are some key social conditions that commonly shape decisions made by producers.

These social and ecological conditions, although often regional or national in scope, interact in a specific location to create a unique set of place-based risks and opportunities. Although farmers and ranchers are experienced risk managers accustomed to navigating the particular details of their operating context, agricultural risk management has traditionally favored using financial and technological assets to reduce specific crop and livestock sensitivities. Managing a farm or ranch with an eye toward enhancing adaptive capacity requires taking a careful look at how *all* resources under management—natural, human, social, physical and financial—might be used to reduce climate risk to operations. For example:

- A healthy natural resource base reduces climate risk by buffering weather variability and extremes. Healthy soils, an adequate supply of clean water, crops and livestock that are well-adapted to local conditions, and diverse habitats that support beneficial organisms and other wildlife are just a few examples of the climate protection services provided by natural resources.

- Effective climate risk management will depend on human capacity, both on the farm or ranch and in the local community, to learn, plan for and adapt to changing climate conditions. Climate risk is expected to make new and different demands on farm management and labor and will likely require a high level of financial and emotional flexibility.

- Support and engagement from the community can reduce climate risks to agricultural operations. High-quality social resources can include experienced mentors, networking groups and loyal customers, as well as the local businesses and public and civic organizations that provide the goods, services and markets important to the success of a business.

- More frequent and intense extreme events are already presenting hazards to critical infrastructure, including buildings used for production and processing, storage structures for feed, forage and farm equipment, and fencing, roads, water and irrigation. As a result, farmers and ranchers are making numerous adjustments in their physical resources—the materials, tools, equipment, technologies and infrastructure they use to manage their production system.

- Financial resources—which typically include family and farm income and assets, insurance, credit and cost-share programs, investments and savings—provide the necessary capital to invest in the other kinds of resources needed to promote high adaptive capacity.

Smart combinations of complementary practices from these five different resource types will enhance adaptive capacity far beyond any one practice used by itself.

Individual Capability

Agriculture is first and foremost about people managing land. Farmers and ranchers are operating on the front lines of climate change. The success of their business depends on their ability to navigate both the ecological and social complexities of an agricultural operation. The adaptive capacity of American agriculture ultimately depends on human capacity, both on and off the farm, to develop new abilities to learn, plan and adapt to changing climate conditions. Taking action to enhance the adaptive capacity of the production system will likely make new and different demands on farm and ranch managers and the people that support them, increase production costs and require a high level of financial and emotional flexibility.[8]

An individual producer's ability to manage the adaptive capacity of their farm or ranch is strongly influenced by their perception of key risks to their operation. Thinking about how best to manage the changing nature of weather-related risks rarely makes it onto most producers' "to do" lists and climate risk is still not commonly addressed in standard agricultural risk management recommendations.[9]

Recent research has documented the ability of rural people engaged in agriculture to detect changes in local climate, such as altered plant and animal phenology, new distributions of species, shorter or longer growing seasons and a shifting frequency of extreme weather events. The changes that people notice tend to be closely related to the aspects of the weather that have the most direct effect on their livelihoods, to such an extent that residents of the same community may identify different changes depending on their occupation. Other factors that influence perceptions of local climate change include recent personal experience of extreme weather events, which can heighten the perception of climate risk, and pre-existing beliefs and attitudes about global warming that lead people to selectively remember weather events in ways that reinforce their worldview. Biased perceptions of local climate changes influence individual capability to act and present barriers to effective climate change adaptation.[10]

The perceptions and concerns of U.S. farmers about climate change issues have only recently begun to be explored. Over the last decade, several large surveys of farmers growing commodity crops like corn, soybeans, rice and wheat with industrial practices found general agreement that weather-related production challenges seemed to be increasing, along with higher insect and disease pressures, but respondents have varying perspectives on the cause of these changes.[11] A number of smaller-scale case studies and survey research with more diversified groups of small and mid-scale producers growing vegetables, grains and livestock reported similar findings. It seems that U.S. farmers and ranchers are making adjustments to changing weather patterns, but many have not yet recognized the unique nature of climate risk as a threat to the sustainability of their operation.

The perception of climate risk, combined with the geographic variability in climate change effects and differences in production system sensitivities to climate effects, all play a role in an individual's motivation to take adaptive action.[12] Ultimately, the success of U.S. agriculture in a changing climate will rest on the willingness and ability of individual producers to take action to reduce climate risk to their operations. In order to take effective action, producers need access to useful knowledge and effective climate-risk management tools to enhance the adaptive capacity of their operations.

Existing Knowledge and Options

Effective management of the production risks associated with weather variability has always been an important consideration in agricultural businesses. There is a large and ever-growing toolbox of management strategies—both old and new—that farmers and ranchers can use to reduce weather-related disruptions to their operations. This knowledge base is an important resource for climate risk management, but not without some rethinking. This is especially true for knowledge and options that have been developed for use in industrial agriculture. This is because industrial risk management favors the application of technologies designed to efficiently target a specific threat associated with a specific sensitivity—for example, that of corn to drought or sheep to worms— and without taking into consideration the farm as a whole. This focus on targeting specific sensitivities without looking at the farm as a whole, already so costly to producers and society, is expected to grow even more expensive and less effective as climate risk intensifies in coming years.[13]

6

Managing Climate Risk: Adaptation Stories

THIS BOOK SHARES the adaptation stories of more than 40 sustainable farmers and ranchers growing food across the U.S. Some are nationally-recognized leaders, others are very happy to stay out of the limelight. Some have been farming more than 50 years, others as little as three. Some are farming land that has been in their family for four or five generations, while others have joined into partnership with their community to be the first in their family to own a farm.

As each farmer and rancher featured in this book tells their story, they paint a rich picture of the whole that they manage. The stories they share are shaped by different experiences and perspectives, advantages and disadvantages, and the hopes and dreams they bring to their work as food producers. Their choices about how best to manage their agricultural business—to decide what to grow and how to grow and market it—are made within an operating context shaped by a global industrial food system that is increasingly challenged by climate change.

One thing that these producers all have in common, no matter where they farm or how much land they manage, no matter their production philosophy or the products they sell, is a different way of thinking about how to grow food. It's a way of thinking about agriculture that goes beyond a simple industrial focus on how to achieve the highest yield out of one or two crops with the latest technology. It's a way of thinking that looks instead at how to achieve their goals through the design and management of a healthy agroecosystem.

These farmers and ranchers produce high quality, nutrient-dense whole foods—vegetables, fruits, nuts, grains, dairy products and meats—throughout the United States (see Figure 6.1). You have already heard from a few of these producers and you will hear from many more throughout the rest of the book. The stories that are shared in full in Part 4 of this book are noted in the regional introductions that follow. Updated stories from the first edition that are not included in Part 4 can be found at realworldresilience.com.

Farmers and Ranchers in the Northwest and Southwest

Nash Huber says that new risks from warmer winters, wetter falls, more unpredictable weather in summer and more variable temperature and moisture extremes have eased some in the last few years as weather patterns have become more like they were 45 years ago when he first started Nash's Organic Produce on the North Olympic Peninsula in Washington. He grows organic vegetables and fruits, food and feed grains, pork, poultry, eggs and cover crop seed for direct markets on about 1,000 acres. Nash has adapted to changing weather and market conditions by purchasing more tractors, tools, combines, processing equipment, shifting from produce to grain production and closing his retail grocery store.

Kole Tonnemaker, along with his son and daughter-in-law, Luke and Amanda, are third and fourth generation fruit growers at the 126-acre Tonnemaker Hill Farm in western Washington near Royal City. Kole credits a shift to higher-value direct markets, certified organic production and the addition of vegetables to the farm's crop mix with their continued success, despite warming winters and more variable seasonal temperatures over the last 20 years.

Russ and Kathy Zenner farmed 2,800 acres of food grains and dry beans in a dryland production system in the Palouse Region of Idaho for more than 40 years. Changes in seasonal rainfall patterns over the last two decades motivated Russ and his cousin Clint—who took over management of the farm with his wife Alicia in 2015—to increase crop di-

FEATURED FARM & RANCH LOCATIONS

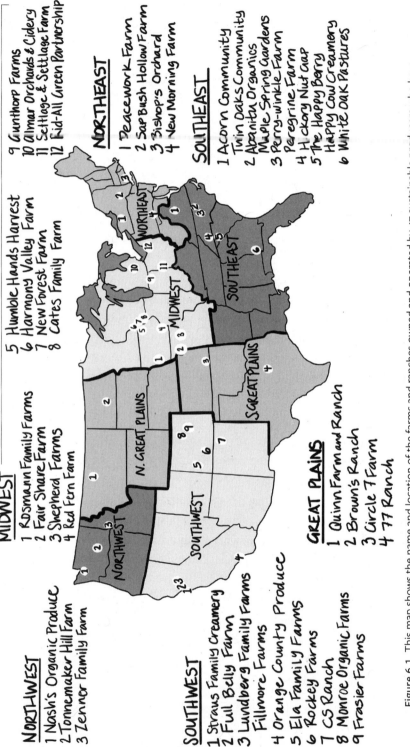

NORTHWEST
1 Nash's Organic Produce
2 Tonnemaker Hill Farm
3 Zenner Family Farm

MIDWEST
1 Kosmann Family Farms
2 Fair Share Farm
3 Shepherd Farms
4 Red Fern Farm
5 Humble Hands Harvest
6 Harmony Valley Farm
7 New Forest Farm
8 Cates Family Farm
9 Gunthorp Farms
10 Almar Orchards & Cidery
11 Settlage & Settlage Farm
12 Kid-All Green Partnership

NORTHEAST
1 Peacework Farm
2 Sap Bush Hollow Farm
3 Bishop's Orchard
4 New Morning Farm

SOUTHEAST
1 Acorn Community
  Twin Oaks Community
2 Abanitu Organics
  Maple Spring Gardens
3 Perry-winkle Farm
  Peregrine Farm
4 Hickory Nut Gap
5 The Happy Berry
  Happy Cow Creamery
6 White Oak Pastures

SOUTHWEST
1 Straus Family Creamery
2 Full Belly Farm
3 Lundberg Family Farms
  Fillmore Farms
4 Orange County Produce
5 Ela Family Farms
6 Rockey Farms
7 CS Ranch
8 Monroe Organic Farms
9 Frasier Farms

GREAT PLAINS
1 Quinn Farm and Ranch
2 Brown's Ranch
3 Circle 7 Farm
4 77 Ranch

Figure 6.1. This map shows the name and location of the farms and ranches owned and operated by 47 sustainable producers who have shared their stories in the first and second editions of *Resilient Agriculture*. You can learn more about these farms at realworldresilience.com.
Credit: Caryn Hanna.

versity and integrate livestock in an effort to enhance soil quality and improve their ability to fine tune the crop rotation no matter the weather.

Albert Straus is the CEO of the Straus Family Creamery and the second generation to manage milk production at the Straus Family Dairy near Tomales Bay, just north of San Francisco, California. A longtime, certified organic, pasture-based dairy producer, Albert says that drought and more weather extremes over the last decade have created new challenges on his ranch and for other pasture-based dairy producers in the area. His focus on creating healthy soils with intensive grazing and compost applications, plus innovative carbon management, has enhanced the resilience of his dairy to a continuing drought in the region (see Part 4).

Paul Mueller co-owns and operates Full Belly Farm, a 400-acre diversified organic farm in the Capay Valley of Northern California where more than 80 different crops including vegetables, herbs, nuts, flowers, fruits, grains and livestock are raised on a diverse landscape mosaic. More heavy rainfall, longer dry periods and continuing drought have encouraged a shift to drought-tolerant cover crops, cover crop mulches, more efficient irrigation and riparian restoration.

Bryce Lundberg and his brother Eric own and manage one of about 40 local farms that produce rice for Lundberg Family Farms in the Sacramento Valley of California near Richvale. Bryce sums up the weather changes that he has noticed over his years growing rice simply as "fewer normal years." To reduce the increased production risks created by more variable seasonal weather patterns, Bryce has shifted to shorter-season rice varieties to have a bit more flexibility in planting and harvest dates and has purchased more field equipment in order to get fieldwork completed when conditions permit (see Part 4).

Ryan Fillmore came back home to manage walnut production full time with his father on their 230-acre family ranch near Gridley, California, about 12 years ago. Despite their efforts to enhance soil quality on the ranch, Ryan says that prolonged drought requires unprecedented post-harvest irrigation to replace winter rains, which have come later or sometimes not at all in the last decade.

A.G. Kawamura is the third generation to grow fresh produce on about 1,000 acres of leased land in and around Orange County, California, to supply Orange County Produce, his family's produce distribution business. His management focus on securing leased parcels with at least two sources of high quality water, utilizing drought- and heat-tolerant cultivars, employing water conditioning and precision agriculture technologies, and swiftly restoring degraded urban soils with cover crops and compost have helped him stay in business despite the prolonged drought, increased labor costs and new competition (see Part 4).

Steve Ela and Regan Choi are organic fruit growers at Ela Family Farms located near Hotchkiss on the Western Slope of Colorado. Steve and Regan manage 100 acres planted with 55 varieties of tree fruits, plus grapes and tomatoes. More variable weather, reduced snowpack, more extreme weather and a lengthening growing season have required a shift to more efficient irrigation systems, changes to more robust fruit species and cultivars, plus the addition of more frost protection. Devastating back-to-back freezes in 2019 and 2020 inspired longtime customers to invest in the farm's recovery.

Brendon Rockey and his brother Sheldon are the third generation to grow potatoes on the 500-acre Rockey Farms located in the San Luis Valley of Colorado. Dwindling water supplies and falling ground water levels motivated Brendon to explore increased crop diversity as a way to reduce water use. His improved potato production system maintains yields, improves crop quality, reduces water use by 50 percent, and reduces or eliminates fertilizer and pesticides through the use of cover, companion and strip crops and integrates custom grazing of cattle and sheep.

Julia Davis Stafford is a fourth-generation co-owner and operator of the CS Ranch, located in northeastern New Mexico on 130,000 acres near Raton. Julia credits holistic management practices like planned grazing and ecological monitoring with the success of the ranch despite nearly 25 years of drought. The continuing drought has required Julia to reduce the ranch's cattle herd by as much as 70 percent in some years, raising concerns about the future of the ranch and nearby communities.

Jacquie and Jerry Monroe, along with their son Kyle, are the third and

fourth generations to own and manage Monroe Organic Farms, a diversified farm in Colorado's Front Range near Greeley. The 200-acre organic
farm produces a hundred different vegetables and all the pasture, hay
and feed grains needed to produce pasture-based meats (beef, pork and
lamb) and eggs on the farm. In response to increased competition for
water, more extreme weather and a longer growing season, the Monroes
have invested in more efficient irrigation, added physical crop protection, and adjusted annual production plans to allow for projected water
availability.

Mark Frasier has managed the 29,000-acre Woodrow division of
Frasier Farms for nearly 40 years. The third generation on the ranch,
Mark buys stocker cattle and raises calves produced on the ranch to run
nearly 5,000 head when fully stocked. Mark has not noticed a change in
weather patterns during his lifetime, but credits planned grazing with
the resilience of his operation to the weather extremes typical of the
short-grass prairie region of eastern Colorado.

Farmers and Ranchers in the Great Plains

Bob Quinn produced certified organic food grains on 4,000 acres in a
full-tillage dryland production system for more than 40 years near Big
Sandy, Montana. More weather extremes and drier fall conditions over
the years required Bob to change his crop mix and field work scheduling,
but warmer winters created new fruit growing opportunities.

Gabe Brown managed the production of cattle, feed and food grains
on 5,000 acres of native rangeland, perennial forages and no-till cropland near Bismarck, North Dakota, for more than 30 years. He points
to high soil quality, crop diversity, adaptive grazing and locally-adapted
livestock as key to the success of his business as flooding rains, temperature extremes and drought increased over the last two decades.

Gail Fuller and his partner, Lynnette Miller, established Circle 7
Farm in 2019 on 160 acres in Severy, Kansas, to produce pasture-based
beef, lamb, pork, poultry and eggs. Gail had owned and operated Fuller
Farms, a 3,500-acre family farm located in east central Kansas near Emporia where he produced a diverse mix of cash crops, cattle, sheep and

poultry in a dryland production system. Increased flooding risk, a heavy debt load and growing success with direct marketing convinced Gail and Lynnette, to sell the family farm, downsize and move to higher ground (see Part 4).

Gary and Sue Price use planned grazing to manage a cow-calf and stocker operation on the 2,500 acres of restored native range and woodlands that dominate the landscape of the 77 Ranch, located near Blooming Grove, Texas, south of Houston. Gary says that healthy soils, native grasslands and restored wetlands support the continued success of their business despite extreme drought and record-breaking rainfalls over the last decade. In 2021, Gary completed the first sale of carbon credits produced on the ranch.

Farmers in the Midwest

Ron and Marie Rosmann, along with their sons, David and Daniel, and Daniel's wife, Ellen, produce certified organic feed and food grains, beef, pork and poultry on the 700-acre Rosmann Family Farms located in west central Iowa near Harlan. Extreme temperature swings require more careful livestock management in winter and a lengthening growing season has created new weed management challenges in annual crops.

Rebecca Graff and Tom Ruggieri manage about ten acres of annual and perennial vegetables and fruits, culinary herbs and a large flock of laying hens at Fair Share Farm in northwest Missouri near Kansas City. Rebecca and Tom agree that extremes of temperature and moisture seem to be more common now than when they started farming about 20 years ago, but they are quick to point out that it is hard to tell because their region is well-known for weather extremes. They have worked for many years to transition the farm to solar power with the goal of reducing the carbon footprint of their business. In 2020, they installed a Masterline design to slow down, capture and store runoff from heavy rains in the soil for use during dry periods (see Part 4).

Dan Shepherd grows, processes and direct markets 300 acres of mature pecans at Shepherd Farms near Clifton Hill in north central

Missouri. Although Dan has not noticed any clear trends in changing weather patterns over his 50 years on the farm, he has experienced several weather-related firsts since 2007: total crop loss from a spring freeze in 2007, a levee breach in 2008 (and again in 2013 and 2019), a prolonged drought from 2011 to 2013, and a total crop loss from hail in 2017.

Tom Wahl and Kathy Dice grow chestnuts, pawpaws, persimmons, heartnuts and Asian pears in a 70-species agroforestry production system at Red Fern Farm in southeast Iowa near Grandview. Since they established the farm 35 years ago, extreme drought and flooding have been long-term weather challenges, along with late spring frosts and early freezes in the fall. The only changes in weather patterns that they have noticed is a cooling of both summer and winter temperatures in the last 20 years.

Hannah Breckbill and Emily Fagan have managed organic vegetables, lamb and pork production at Humble Hands Harvest in northeast Iowa near Decorah since 2016. Hannah has been a vegetable grower in the region for about a decade and in that short time has suffered total crop losses from both record-breaking flooding and drought. Cooler and wetter conditions in the spring have made staying on a planting schedule difficult, so Hannah and Emily added protected growing space and have been experimenting with no-till, raised beds and silage tarps to manage weeds before planting and allow for timely spring planting of field-grown crops (see Part 4).

Richard DeWilde owns Harmony Valley Farm, a 200-acre diversified vegetable farm located near Viroqua in southern Wisconsin. He realized that there "is no normal anymore" after back-to-back, 2000-year flood events in 2007 and 2008 caused catastrophic flood damage. Continued flooding, heavier winter rains and cold, wet spring weather over the last decade have increasingly disrupted spring planting, created new nutrient management challenges and required relocating vegetable production to higher ground.

Mark Shepard has managed the production of certified organic vegetables, fruits, nuts, cattle, hogs and nursery stock at New Forest Farm in central Wisconsin near Viroqua for more than 25 years. Mark appreciates how regional marketing networks along with soil health and crop diversity in his "teenaged" agroforestry system sustain the profitability of his farm as increases in flooding rains, dry periods, drought and extreme weather create new risks for agricultural producers in his area (see Part 4).

Dick and Kim Cates, along with their son Erik and his wife Kiley, own and operate Cates Family Farm, a 900-acre grass-finished beef operation near Spring Green in southern Wisconsin. Over the last decade, the family has worked to manage new risks from increased winter rainfall, more frequent flooding and extreme high and low temperatures in summer and winter by adding stream crossings, leaving more pasture cover after grazing or mowing, and moving cattle inside during extremes of cold weather.

Greg and Lei Gunthorp are the fourth generation to produce pasture-based cattle, swine and poultry at Gunthorp Farms near LaGrange, Indiana. Livestock produced on the farm are slaughtered, processed and packaged on-farm for regional distribution to direct wholesale and retail markets. Greg can't say that he has noticed any significant changes in weather during the nearly 40 years he has managed the farm, except for possibly a slight increase in extreme weather events.

Jim Koan and his wife Karen, their son Zachary and daughter Monique own and operate Almar Orchards and Cidery, a fifth-generation diversified organic farm on 300 acres located in eastern Michigan near Flint. They produce apples, pears, pasture-raised hogs, maple syrup and a diverse line of drinking vinegars and hard ciders sold online and at an on-farm store and tasting room. Jim added on-farm processing to farm operations about a decade ago to overcome new risks and declining profits created by changing weather patterns, especially more frequent late spring frost/freeze and heavy rainfalls, plus increased competition from low-priced imported fruits.

Jordan Settlage transitioned his family's 500-acre farm in northwest Ohio from conventional grain production to organic, pasture-based dairy just after he graduated from college in 2014. Jordan appreciates the resilience benefits of his decision to make the shift from row crops to perennial forages as weather extremes—particularly flooding—become more frequent and intense at Settlage and Settlage Farm (see Part 4).

Marc White, Keymah Durden and Dave Hester have managed different aspects of Rid-All over the years, from the production to the marketing of compost, bedding plants, fresh vegetables, fish, medicinal herbs and healthy beverages at the Partnership's urban farm in Cleveland Ohio. Even though much of the farm's production is under cover, seasonal production losses caused by more frequent and intense weather extremes have motivated some changes over the years, including shifting to higher-insulation plastic to cover indoor growing spaces, upgrading irrigation systems and investing in backup power generators (see Part 4).

Farmers in the Northeast and Southeast

Elizabeth Henderson grew organic vegetables at Peacework Farm for more than 25 years near Newark in upstate New York. Hotter summers, more heavy rainfall and drought, and novel diseases required adjustments to the farm's management practices, including the addition of irrigation, changes to field work hours and adjustments to crop succession timing.

Jim and Adele Hayes have raised beef, lamb, pork and poultry on 160 acres of pasture at Sap Bush Hollow Farm west of Albany, New York, for more than 40 years. More dry periods and drought, more frequent and stronger winds and extreme weather have required the addition of new infrastructure such as drainage systems, farm ponds, a raised barn, reinforced pasture shelters and solar power.

Jonathan Bishop is a fourth generation co-owner and CEO at Bishop's Orchards, a 320-acre diversified fruit and vegetable operation with a full service grocery and winery in Guilford, Connecticut. Although Jonathan hasn't noticed any long-term changes in weather patterns, a number of severe storms over the last decade have confirmed for him the resilience benefits of scale, experience and crop diversity in agricultural businesses.

Jim Crawford produced organic vegetables and small fruits on 45 acres at New Morning Farm in south central Pennsylvania for nearly 50 years. As more variable spring weather, more heavy rains and summer droughts, and a growing number of novel plant diseases began to complicate vegetable and fruit production on the farm, Jim shifted soil preparation to the fall, physically protected soils and crops from excessive rainfall with plastic, moved crop production out of the floodplains, and for the first time as an organic grower, began using OMRI-approved pesticides.

Ira Wallace and Mary Berry grow vegetable seed on 4.5 acres and manage processing and marketing seed sourced from a network of about 70 small- and medium-scale farmers for Southern Exposure Seed Exchange, a cooperative business owned by the Acorn Community in central Virginia, near Mineral. Crop failures from heavy rains and flooding have increased in recent years at Acorn and in their growers' network. They are adapting to these new climate risks by shifting to no-till practices to increase soil health, expanding protected growing and seed processing space, and are making plans to install some earthworks to improve drainage (see Part 4).

Pam Dawling has grown vegetables and fruits on about 3.5 acres for Twin Oaks, an intentional community located in central Virginia near Louisa, for about 20 years. Pam thinks that heavy rains, dry periods and droughts, and extreme temperature swings have become more common in recent years. She's adjusted to these changes by focusing on getting field work done whenever conditions permit and expanding protected growing space (see Part 4).

C. Bernard Obie, known as "Obie" to his friends, grows biodynamic vegetables and fruits on about 12 acres in central North Carolina at Abanitu Organics near Roxboro. Obie says that heat waves, damaging storms and heavy rains have been consistent challenges over the last 15 years or so that he has been growing full-time at Abanitu. He manages these production risks with adaptive planting, physical protection, raised beds and careful attention to irrigation (see Part 4).

Ken Dawson has produced organic vegetables, small fruits, culinary and medicinal herbs, and bedding plants on 80 acres at Maple Spring Gardens, located in central North Carolina near Durham for more than 40 years. Over the last decade, Ken and his daughter Sunshine have managed new risks associated with increasing temperatures and a series of recording-breaking weather extremes by growing in shade structures in summer, adjusting planting schedules to avoid new disease challenges, shifting to no-till planting when necessary and moving to year-round production and direct marketing with a focus on high-value, cool-season crops.

Cathy Jones and Michael Perry grow organic vegetables, pastured poultry and cut flowers on ten acres at Perry-winkle Farm in central North Carolina near Chapel Hill. When they started farming 30 years ago, they decided to avoid plastic in their operations, but more dry periods, drought and heavy rains have got them questioning this long-held principle. They have adapted by tightening up their rotations to keep the soil covered by crops or cover crops at all times and preparing raised beds for planting throughout the year whenever soil conditions are appropriate.

Alex and Betsy Hitt managed the production of vegetables and flowers for wholesale and direct markets at Peregrine Farm south of Chapel Hill, North Carolina, for more than 40 years. More intense heat waves and drought in summer, combined with a reduced summer water supply, got them thinking about adding water storage capacity and shifting crop production away from summer and to fall, winter and spring.

Jamie and Amy Ager own and operate Hickory Nut Gap Meats and manage the production of regenerative beef, swine and poultry on the Ager family farm just east of Asheville, North Carolina. Jamie says that managing weather variability is just a normal part of farming, but in the past few years more frequent heavy rainfalls have started to get his attention (see Part 4).

Walker Miller and his family grow about 14 acres of small fruits—blueberries, blackberries, muscadine grapes, figs, elderberries and persimmons—at the Happy Berry Farm in upstate South Carolina near Clemson. Over the last decade, Walker has shifted to new cultivars, adopted new soil health practices and planted shade trees to reduce increasing production risks created by new pests and diseases, warming temperatures and earlier spring warmup, increased frost/freeze risk, longer dry periods and more intense drought (see Part 4).

Tom Trantham owned and operated Happy Cow Creamery, a 90-cow grass-based dairy farm and creamery located south of Greenville, South Carolina, for more than 35 years. Drier summers and more intense summer thunderstorms over the last decade have not presented any challenges on the farm, largely because Tom's use of a dynamic rotation of diverse annual forage crops planted into continuous pasture makes it easy to adjust to changing weather conditions throughout the year.

Will Harris and his daughters, Jenni and Jodi, own and operate White Oak Pastures, a fifth-generation regenerative livestock farm located in southwest Georgia near Columbus. In the last five years, rainfalls have grown heavier and dry periods are more frequent and last longer. A direct hit from Hurricane Michael in 2018—the first hurricane to hit the farm in all of its 154 years—caused $1.4 million in damages. Will credits diversified operations, dynamic pasture management, on-farm processing and direct marketing for the swift recovery and continued success of his business.

A New Path for American Agriculture?

Exploring the unprecedented challenges of climate risk in terms of climate vulnerability—exposure, sensitivity and adaptive capacity—has encouraged me to think about managing risk in food and farming in an entirely new way. I am no longer satisfied with risk management solutions that focus on just one element of the farming system with little regard for how that solution will affect relationships both on the farm and beyond. I am astonished by our willingness to accept the extraordinary costs—in natural, human and social resources—of maintaining the adaptive capacity of agriculture through financial and technological investments. I am outraged by public policy designed to protect "business as usual" despite all that we now know about the many ways that industrial thinking does damage to land, people and community.

Let me explain.

Broadly speaking, the adaptive capacity of industrial models of production arises from the management of large land holdings (typically, owned plus rented land), purchased inputs (e.g., irrigation, fertilizer, pesticides) and government subsidies (e.g., for agricultural research and development, technical and financial assistance programs, direct payments and insurance, agricultural labor exemptions, disaster payments, food security payments, public health payments) to produce commodity products without regard for local resource conditions.

In contrast, the adaptive capacity of sustainable models of production arises from the management of smaller land holdings (typically owned), production inputs produced by healthy soils and agrobiodiversity (e.g., capture, storage and release of natural precipitation, crop nutrients released by decomposition, pest suppression by beneficial insects) and social capital (e.g., direct markets, community-based research and education) to produce high-value food products that are well adapted to local resource conditions. The choices that agricultural producers routinely make about the assets they manage—people, land, crops, livestock, infrastructure and finances—determine in large part the ability of the farm or ranch to sustain production under challenging climate conditions.

Over the last 50 years or so, industrial agriculturalists (and all the people that support them) have emphasized technological and financial assets to enhance the adaptive capacity of their farms and ranches both to weather and to market disturbances and shocks, while sustainable agriculturalists (and all the people that support them) have emphasized natural, human and social assets to do the same. While each of these different strategies offers some climate risk management benefits, the adaptive capacity of both industrial and sustainable production systems could be enhanced by taking advantage of all asset types and a greater diversity of options within each asset type.[1]

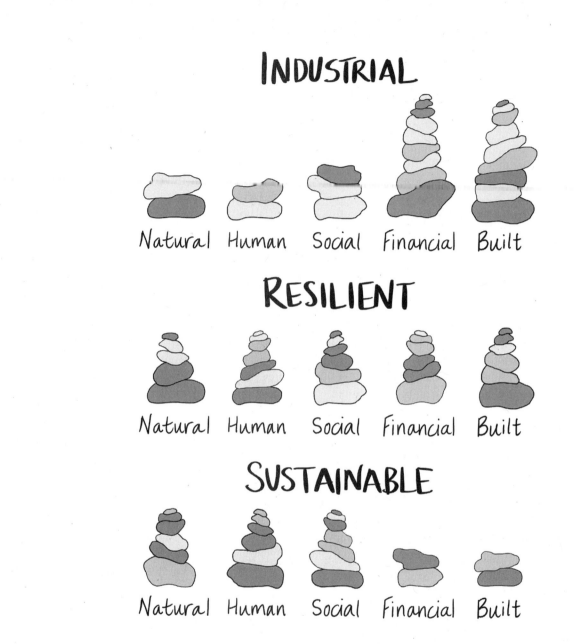

Figure 6.2. Industrial and sustainable producers tend to rely on different kinds of assets to manage the fragility of their operations. Because industrial operations are subsidized by the public, they rely on financial and technical assets to recover from disruptions. Because sustainable producers do not have access to these subsidies, they rely on natural, human and social assets to reduce the fragility of their operations. The adaptive capacity of both kinds of operations would be enhanced by access to the full range of high-quality assets provided by the resilient management portfolio. Credit: Caryn Hanna.

One way to think about this is to notice that industrial agriculture is ignoring the wisdom in the old saying "Don't put all your eggs in one basket." Industrial agriculture, and the people that support it, are betting for all of us that the solutions to our current predicament reside in money and technology. But sustainable agriculture is doing something similar— betting that the solutions reside, more or less, depending on the brand of sustainable, in biodiversity, ecological management and local food.

Both production philosophies hold pieces of the climate risk management puzzle. I wonder if climate change is the challenge that finally begins to erode the longstanding political divide about the "right" way to grow our food? Will climate change be the crucible that forces us all to question long-held beliefs about the purpose of agriculture as we forge a new way to think about our foodways? Could confronting climate change together help us find, not common ground, but new ground rooted in respect and appreciation for all that we have learned—as foragers and farmers over the last 10,000 years—about how to cultivate healthy relationships with the land, people and communities that feed us?

I see glimpses of this new story emerging across America and beyond. A recent USDA adaptation guide[2] recommends that all producers, regardless of production philosophy, put a priority on improving natural resource quality on their farms and ranches. Climate-smart producers are using technological solutions to reduce the damages to land, people and community generated by industrial practices such as large-scale monocrops, the use of synthetic pesticides and fertilizers, and confined animal feeding operations. Organic farmers are leaning more heavily on organic pesticides and fertilizers. Regenerative farmers are adopting soil conservation practices used for decades by sustainable farmers. Contemporary land managers of all kinds are looking to Indigenous cultures for new ways to think about ecological foodways. And everybody is talking about resilience.

PART 2

THE RULES OF RESILIENCE?

When I started farming there was a pattern to the weather,
one year was kind of like the one before.
I think it's more challenging to be a farmer now.
No two years are alike. You have to be so nimble these days.

— Elizabeth Henderson, Peacework Farm, Newark, New York

A New Way to Think
About Solutions

FLASH DROUGHT. New pests and diseases. Hot summer nights. Empty wells. Flooding rains. As difficult as our changing weather is for farmers and ranchers, climate risk does not usually make it to the top of their list of worries. Weather generally is pretty low on a list that usually has labor, markets, profits and regulations jockeying for a top position. Climate risk is often seen as the least of their worries.

The truth of the matter is that climate change is just one of a whole host of unprecedented challenges that the people who feed us are facing in these times. A failing Colorado River threatens the water supply of 40 million Americans. Young farmers compete for land with global investors looking for a safe haven and a lucrative return. American producers struggle to stay in business as the industrial food system drives a global race to the bottom. And in the spring of 2020, all the people who feed us—growers, farm and food processing workers, and the workers in food retailing who serve us in restaurants and grocery stores found themselves suddenly standing on the front lines of a global pandemic.

The way these challenges come together in our food system reveals the fragility of a way of life built on three hundred years of relentless exploitation of land, people and community. With the growing awareness that "business as usual" is no longer an option, a multitude of silver-bullet food and farming solutions are bubbling up all around us. The question is not *whether* we change, but *how* we will change the way we eat. Although it may be comforting to imagine, it is pretty clear that simply changing the brand we buy is not going to get the job done. The soil will not save us, we do not live on a vegan planet and many selling the regenerative ag solution seem to have forgotten that the people who feed us and the communities they call home are in as much need of healing as the soil.

So many different solutions—how do we choose among them? Some represent relatively minor tweaks to the existing food system, while others require fundamental change. And every solution seems to focus on a different part of the food system. Some see the answer in carbon markets, others by eating less meat, some advocate growing food in city skyscrapers and others imagine a world in which food is printed on demand.

The problem is that none of these solutions address the whole picture. It's like that fable about the six blind men and the elephant, each studying a different part of the animal, and then disagreeing on their findings. This story is a cautionary tale told in many different cultures. It is deep wisdom from our ancestors that warns us to beware of the limitations of any one perspective. It is a lesson that invites us to remember that truth resides in a collective vision. So many food climate solutions suffer from this problem of limited perspective—each focusing on just one part of the food system without considering the whole.

As I explored the noisy world of food and farming climate solutions, I realized that we need a better way to think about the massive global networks that feed us. We need a way to keep this big picture in mind even as we focused in on a specific part to understand it better. We need new words, a new perspective and new strategies to help us leave behind the intense political tribalism of our times. To find a way to come together and imagine a food system capable of sustaining land, people and community through the inevitable challenges ahead. We need a way of thinking that inspires everyone to take action to bring this shared vision into being.

It was in this search for a new way of thinking that I ran into resilience. In those early days, resilience thinking was pretty new and not many people were doing it. Of course, since then, interest in resilience has exploded as we have been continually battered over the last decade by climate chaos, economic meltdowns and, most recently, a global pandemic that seems to have no end.

These days, everybody is talking about resilience, but what do they really mean? Most people have in mind resilience concepts developed in psychology, but the ideas that I share in this book come from social-ecological resilience science. Social-ecological resilience science is a branch of ecology that specializes in understanding the behavior of ecosystems dominated by humans—either intentionally, for the purpose of producing goods or services, or unintentionally, through the collateral damage caused by human activity.

As I started learning more about social-ecological resilience thinking, I began to get excited. I started to see that it could help us overcome many of the limitations associated with the way we make decisions about food as a climate change solution.

Resilience thinking is uniquely suited to innovating climate solutions for at least three reasons:

- It is a *systems science*,[1] so it offers a useful set of concepts and language that can *improve our ability to understand complex change* in situations which are difficult to define, much less improve.
- It is an *applied science*, so it offers *practical tools to help us evaluate* how actions taken today are likely to influence progress towards our goals.
- It is *grounded in ecology*, so it invites us to keep in mind *the realities of how our planet actually works* and our place within it.

We are living in a time of climate change. We know that in coming years we are going to be challenged by more frequent and damaging weather-related disturbances and shocks. That's really clear. And we know that these disturbances and shocks are going to damage much more than agriculture. We're already seeing the effects of climate chaos throughout the world and increasingly in the places that we call home. Every time there is a damaging weather event, we have a choice. We can choose to bounce back to twentieth century industrial standards, or we can use social-ecological resilience thinking to bounce forward to twenty-first century resilience standards that put to work everything we've learned in the last century about how to live well on this planet.

Resilience Is Not What You Think

The concept of resilience has origins in a diverse set of disciplines, including engineering, ecology, psychology, human health and disaster management.[2] Each of these disciplines brings a different approach to resilience that has bearing on the way I think about resilient agriculture. For example, concepts of resilience within engineering focus on designing systems for tolerance to a predetermined range of disturbance or stress. This type of resilience, called robustness, is certainly useful in situations when the threats to the system and the system's response to those threats can be reliably predicted. But I wondered: Is robustness even relevant to living systems—especially given the uncertainty of climate change threats?

When people say that they are going to "build" resilience, that they want to "build back better," or that they are going to "bounce back" from a damaging shock or disturbance, I wonder if they realize that they are drawing on engineering resilience concepts that are most appropriate for non-living, human-engineered systems like buildings, roads, and barriers to wind and water.

The resilience science developed in ecology, psychology and disaster management are more relevant to food and farming, because these disciplines have developed methods to assess, monitor and manage resilience as a dynamic quality of complex living systems with the ability to respond and adapt to change.[3] When people say that they are going to "cultivate" resilience, that they want to "increase capacity for response" to more extreme weather, or that they want to "bounce forward" from a damaging shock or disturbance, they may not realize it, but they are drawing on resilience concepts developed through the study of complex living systems such as ecosystems, people and communities.

The resilience science developed by social-ecologists has a long history of application in natural resource management.[4] More recently, social-ecological resilience science has been applied to climate change vulnerability and adaptation in many types of human-dominated systems,[5] for example cities,[6] neighborhoods[7] and landscapes,[8] as well as agriculture and food.[9] This kind of resilience thinking gives us a new language to explore complex living systems from different vantage points and integrate different perspectives into a new framework for understanding and managing the qualities and behaviors likely to sustain living systems over time.

Describing a Social-Ecological System: Focal Scale, Identity and Desirability

Resilience thinking uses ecological language to enhance our ability to describe complex agroecosystems[10] in a way that helps us keep in mind both the parts and the relationships between the parts while we explore how the system responds to changing conditions. It also helps us describe the system from different vantage points, zooming out to see the system as a whole, or focusing down on a specific part of the system.

This awareness of *scale* invites us to always keep in mind that there are conditions influencing the behavior of the system at scales just above and below the scale of immediate interest, which is known as the focal scale (because this scale is the focus of interest). For example, when exploring weed response to field management practices, it is often helpful to consider how soil conditions (the scale just below the field scale) as well as the farm landscape (the scale just above the field scale) might influence the success of a particular practice.

Changing weather patterns have increased weed management challenges at the Rosmann Family Farms in Iowa. Over the last two decades, Ron Rosmann has watched giant ragweed slowly move uphill from wetter low-lying areas on the farm, where he can manage it effectively with cultivation, into grain production fields. "We use a ridge-till system, which is a minimal tillage system. We don't do any spring tillage until we are ready to plant. We historically are not even thinking about planting corn until May 1, but now the giant ragweed is already getting big by then. It is a tough weed to take out with cultivation." Ron thinks the combination of wetter weather in May and June, an earlier spring warm-up and a longer growing season have worked together to create new weed management risks in his grain operation.

❧❦❧❦❦

Identity refers to the characteristic structure (the parts), function (the relationship between the parts) and purpose (the goals) of a system. For example, pasture-finished beef and concentrated cattle feeding operations are very different in structure, function and purpose, giving each a distinct identity that is easily recognized. Farming operations that fall within a distinct identity will likely have some subtle differences, but they will be consistent in structure, function and purpose. Describing farming systems in terms of identity provides a useful framework for objective comparison and contrast—one that helps us set aside individual beliefs, values and attitudes and to integrate different perspectives of the system.

For example, most pasture-based beef operations will be similar in *structure* (a mixed age group herd is managed to produce calves and finished cattle on pasture), *function* (energy flows into the system from the sun to pastures to cattle and is exported as beef, and *purpose* (production of beef for specialized, value-added markets). Most concentrated cattle feeding operations also share a similar structure (cattle of uniform age are produced on bare dirt lots and fed imported grain), function (fossil/solar energy, water, nutrients imported as cattle and feed, exported as cattle and wastes) and purpose (land- and labor-efficient production of beef for low-value commodity markets).

Resilience thinking makes it easier to consider important questions about the *desirability* of the system, in other words, the ability of the system to achieve environmental, social and economic goals as defined by individuals and society. Describing food and farming systems in terms of scale, identity and purpose invites us all to acknowledge the underlying assumptions that are at the heart of the many conflicting perspectives about how to feed ourselves in a changing climate.

Taking the time to step back and objectively examine the desirability of the system brings up important questions critical to the assessment of sustainability and resilience: Is the system achieving its stated goals? Are there other ways to structure the system to improve performance? Can system functions be changed to reduce ecological and social costs or to increase benefits?

After years of controversy over his use of cover crops cost him his line of credit at the bank, and neighboring land owners canceled land leases because of his "risky" farm practices, Gail Fuller decided to take a step back and reconsider his reason for farming. He and his partner Lynnette spent some time looking at the kind of farming that would best help them reach their big-picture goals. They wondered if continuing to farm 3,200 acres of grain and livestock at Fuller Farms was going to get them where they wanted to go. "When I stopped and really thought about it, I realized that I was farming too much land. Trying to farm that many acres without any credit is impossible. I realized that I was just digging myself deeper into what was already a pretty deep hole. It wasn't long after that Lynnette and I realized that downsizing the farm and getting more into direct marketing might be the solution. We didn't really set a goal for how small we wanted to get. I don't think we really knew, but by 2016 we were managing about 400 acres of my home farm." As Gail and Lynnette gained experience producing grain and livestock products for direct markets, a heavy debt load and increasing flood risk got them thinking about the potential benefits of selling the family farm and relocating to step down to an even smaller farm. After years of discussion, the Fuller family decided to sell the farm and in the fall of 2019, Gail and Lynnette moved to their new farm about an hour south of Emporia near Severy. Circle 7 Farm is about 160 acres of diverse uplands that includes native grasslands, some annual croplands, six ponds, an orchard and a purpose-built event center with a commercial kitchen.

૭◯◯૭ ◯◯◯

Resilience thinking also encourages us to explore plausible current or future events that present critical threats to the identity of the current system. If the system is performing well, the goal of management might well be to enhance the adaptive capacity of the current system in anticipation of the increasing risks associated with expected changes in weather. If the system is not fulfilling its purpose, resilience thinking can be used to transition the system into a new, more desirable identity that is also more resilient to changing climate conditions. Resilience thinking provides the practical tools needed to take a step back from the system, to see it with new eyes and to question underlying assumptions and rationalizations. This is resilience thinking's transformative superpower!

More Than Bouncing Back

Social-ecological resilience thinking can help us go beyond the old "bounce back" idea used by engineers and dig deeper into the full range of options available for cultivating resilience in food and farming. The resilience of living systems is determined by three complementary capacities that work in different ways to sustain the system over time by protecting the system from damaging change, recovering from damage caused by change (the bounce back) and transitioning the system to a new identity when necessary or desired (the bounce forward). Resilient systems tend to cultivate all three capacities.

- *Response capacity* is the ability to respond to disturbances and shocks—both those that are anticipated and those that are not—in ways that avoid or reduce damage and to pivot swiftly to take advantage of new opportunities. Practices that enhance response capacity involve cultivating system relationships that can eliminate damages associated with disturbances, so there is never a need to bounce back. When there is damage, rather than "bouncing back" to an outdated twentieth century standard, we can instead make investments that improve response capacity.

- *Recovery capacity* is the ability to return to normal function swiftly and at low cost in the event of a damaging disturbance. This idea of "bouncing back" is usually what people mean when they talk about resilience. While bouncing back from damage is important, relying on recovery capacity to sustain a system as conditions change tends to become more costly and less effective over time.

- *Transformation capacity* is the ability of the system to make fundamental changes in order to enhance response and recovery capacity as conditions change. Managing for transformation capacity is uniquely valuable as we wake up to the need to transform agriculture and food systems.

Once you know about these three complementary capacities, you can begin to appreciate just how limited the old "bounce back" thinking is. Social-ecological resilience thinking opens up a whole new world of possibility perfectly suited to the practical realities of managing our businesses and communities to thrive in changing conditions. Let's take a closer look at how some sustainable producers in the U.S. cultivate response, recovery and transformation capacity on their farms and ranches.

Response

Agricultural practices that promote response capacity increase the ability of the farm or ranch to continue to function across a wide range of conditions. These practices include relatively simple adjustments to the production system such as replacing an existing crop variety with a more

robust variety, adding mulch to conserve soil moisture and reduce soil temperatures, or adding field irrigation or drainage systems. More complex adjustments involve diversifying the existing crop production system by increasing the biodiversity of the system, for example by adding cover crops, new annual or perennial crop species, or livestock. Adjustments that enhance the response capacity of a farm or ranch can also be made in non-crop production areas, for example by restoring wetlands and riparian areas.

These kinds of diversification practices are important to help maintain production in more variable weather and more frequent and intense weather extremes. These practices achieve these aims by assuring the supply of critical resources needed for crop growth and development, or by spreading production risks through the season by including crops that differ in their sensitivity to specific weather-related disturbances. Diversification can also reduce production costs, because the biodiversity in the system can promote ecosystem services that reduce or eliminate the need for inputs such as fertilizers, pesticides and irrigation. Ecosystem restoration can reduce soil erosion and flooding both on the farm and in communities downstream.

Along with using targeted practices to manage specific known risks, cultivating response capacity can also involve using practices that enhance the general resilience of the cropping system by reducing the risk of damage from unexpected disturbances and shocks. For example, many farmers and ranchers who operate highly diversified systems appreciate the flexibility that biodiversity gives them when it comes time to make management decisions based on actual or projected weather conditions.

Gabe Brown uses practices like dynamic crop rotation and cover crop cocktails to enhance the resilience of his operations to more variable weather and extremes. Planting a diverse mix of crops throughout the growing season gives him the ability to fine-tune his crop rotation plan to adjust to current weather conditions. "That's the beauty of the diverse system of ours," Brown explains, "At times, we want to plant the cover crop and then if the weather conditions change, maybe it's dry, we'll change the mix of species a bit for species that can handle drier conditions, or vice versa. It just makes management so much easier."

C. Bernard Obie says heavy clay soils make the production risks created by heavy rainfalls particularly difficult to manage at Abanitu Organics. "A heavy rain presents an issue with getting into the field to work," says Obie. "The soils just don't dry out quickly and there is a tendency for plants to be stunted or even killed in low-lying areas because moisture sticks around so long in our soils." Obie built a set of raised beds to improve his ability to adapt to wet soil conditions. "In a year where there's a lot of rain and it's early and the temps are cold, the raised beds give us some additional options," says Obie. He also notes that the challenge of farming in clay is a good example of the "give and take" of farming, because although clay soils are slow to dry out, they also provide some welcome drought tolerance.

ᦒᦒᦒᦒ ᦒᦒᦒ

Crop and livestock diversification go hand in hand with product and marketing diversification, which adds another dimension to response capacity. Spreading marketing risk among different types of markets— such as direct sales to consumers at farmers markets and direct wholesale to custom food processors and restaurants—can buffer an operation against shocks.

Focusing on producing high-value products or selling into high-value markets can also reduce risk and increase profitability. Because high-value direct and direct-wholesale markets often welcome uncommon and value-added products, they present a financial incentive for diversifying crop production. In addition, these kinds of markets offer producers the opportunity to use robust crop cultivars and livestock breeds that may not meet commodity market standards, but are well-suited to regional growing conditions.

Recovery

Holding some critical resources in reserve in case of damage is key to enhancing the recovery capacity of agroecosystems. Recovery reserves can be natural (storing extra water and feed supplies), human (experience and ease of managing loss and change), social (community memory, knowledge, skills, public assistance), financial (insurance and disaster payments, savings, access to capital) or physical (backup and alternative energy sources, storage, shelf-stable products). Resilient farms and ranches accumulate reserves across the full range of resources under management.

 Jim Koan appreciates the way that recent weather challenges have forced him to think outside the box, anticipate what could go wrong and plan for the worst. The Flushing, Michigan, apple grower began on-farm production of hard cider—JK Scrumpy's—in part to enhance the recovery capacity of his operation. By processing a portion of his apples into hard cider in years when yields are high, he has a nonperishable product he can sell in years when weather disrupts production. This helps him maintain his cash flow. "In 2012, we had only had ten percent of our apple crop. I had half a million dollars invested in those apples," Koan said. "That was not as big an issue for me as it would have been if we hadn't had JK Scrumpy's. I can walk away comfortably saying that I actually made a profit in 2012."

"In August of 2007 we got hit really hard with some really weird flooding caused by 18 inches of rain in a less than a 24-hour period," Wisconsin vegetable farmer Richard DeWilde recalls. "A lot of crops were peaking just then, like tomatoes. We had pretty big losses because a lot of our farm land is along the Bad Ax River. They called that a thousand-year event. And then we had another one nine months later. That was when I said, 'There's no such thing as normal anymore.' Not many people understand that vegetable farmers have little to no insurance against weather. We can participate in the USDA's NAP program and we do. N-A-P is the abbreviation for Noninsured Agricultural Production. Noninsured meaning it's not corn, soybeans, cotton, wheat. It's not a commodity. NAP is a poor program. It's totally inadequate and we really don't have much else. But if you're a corn farmer, you can buy government-supported, 90-percent-guaranteed income on the corn crop. It's gross. We should care more about feeding people than raising corn for export and ethanol and corn syrup."

Transformation

Ultimately, the resilience of an agricultural business depends on taking a long view of the operation as a whole. What kinds of more fundamental changes in production system practices, enterprises and markets might be made to enhance the overall resilience of the business now and to prepare for the future? Examples of transformative changes that enhance resilience to more variable weather and extremes could involve shifting from monoculture to diversified crop production or from annual to perennial crops, transitioning from housed to pasture-based livestock production or extending farm operations to include value-added processing.

Tom Trantham's pasture-based dairy system boasts many resilient characteristics, but it was not always that way. Like many American farmers feeling the pain of consolidation in the agricultural sector during the 1980s, Trantham was producing a lot of milk in a conventional operation but barely turning a profit. "I went through some really rough times in those days. We all did," he recalled. Although he had long been among the top milk producers in his state, rising feed and farm chemical costs, and falling prices left him with few options when he was refused an operating loan in 1987. After some research into intensive grazing practices, he successfully guided the transition of his 90-cow dairy from feed-based to the forage-based production system he continues to use today. He has dramatically lowered his costs while increasing herd health, milk quality and soil quality. With the opening of the Happy Cow Creamery in 2002, Trantham's transformation from commodity dairyman to specialty milk retailer was complete. Whole

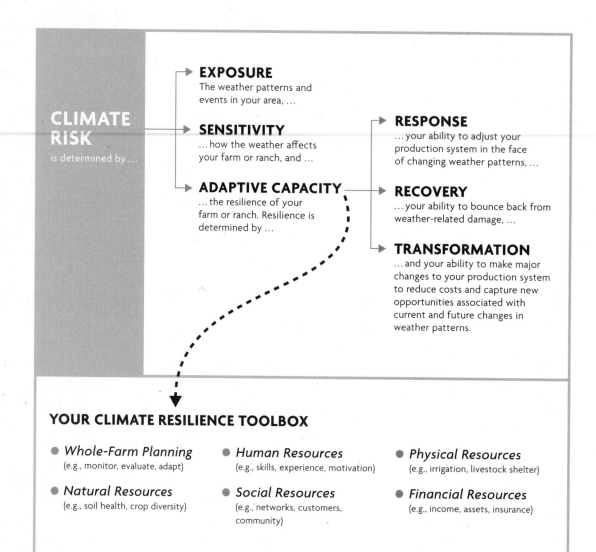

CLIMATE RISK
is determined by ...

EXPOSURE
The weather patterns and events in your area, ...

SENSITIVITY
...how the weather affects your farm or ranch, and ...

ADAPTIVE CAPACITY
...the resilience of your farm or ranch. Resilience is determined by ...

RESPONSE
...your ability to adjust your production system in the face of changing weather patterns, ...

RECOVERY
...your ability to bounce back from weather-related damage, ...

TRANSFORMATION
...and your ability to make major changes to your production system to reduce costs and capture new opportunities associated with current and future changes in weather patterns.

YOUR CLIMATE RESILIENCE TOOLBOX

- *Whole-Farm Planning*
  (e.g., monitor, evaluate, adapt)

- *Natural Resources*
  (e.g., soil health, crop diversity)

- *Human Resources*
  (e.g., skills, experience, motivation)

- *Social Resources*
  (e.g., networks, customers, community)

- *Physical Resources*
  (e.g., irrigation, livestock shelter)

- *Financial Resources*
  (e.g., income, assets, insurance)

Figure 7.1. Understanding the vulnerability of an agricultural operation and options for cultivating response, recovery and transformation capacity is the key to managing climate risk. Resilience thinking can improve climate risk management by clarifying how the full range of assets under management can be used to enhance adaptive capacity. Credit: USDA.[11]

milk, chocolate milk, buttermilk and eggnog are processed on-farm and sold into regional direct-wholesale markets and at an on-farm store that also sells a diverse line of mostly locally sourced fresh and processed products including vegetables, fruits, butter, cheeses and meats.

Colorado fruit grower Steve Ela credits a shift to direct marketing strategies as the reason he has remained in business in a region that has lost 75 percent of its fruit growers over the last 20 years. His family farm was on the same downward path when he took over management in 1990. With declining profits taking a toll on family finances, he looked to direct marketing to turn the farm around. "We started making changes because of bad economics and now we direct market 100 percent of our fruit," Ela said. "We've completely changed our business model in 12 years. Fortunately it's worked, we're still here." Direct markets have also opened up new opportunities for him to diversify crops, because his customers are willing, even eager, to try something new.

∾◌◌∾ ◌◌◯

A good time to consider transformative changes such as those made by Trantham and Ela is when it looks like the costs—ecological, social or economic—of maintaining the current production system are growing unacceptably high, the performance of the production system begins to consistently fall short of meeting goals or it becomes clear that change is inevitable because of growing environmental, social or economic challenges.

The Qualities and Behaviors
of Resilient Systems

R ESILIENT SYSTEMS come in all shapes, kinds and sizes. They can be a natural ecosystem, like a forest, a prairie or a wetland, or they can be social-ecological systems such as a family home, a rural neighborhood, a farm, a ranch or a city. No matter the scale or extent of human influence, all resilient systems depend on some key ecological and social behaviors for their well-being over the long term. These behaviors tend to produce the kind and quality of resources that cultivate high response, recovery and transformation capacity in both natural and human-dominated ecosystems. In natural ecosystems, these behaviors evolved over time as place-based expressions of natural resource conditions. In human-dominated ecosystems, these behaviors can be thought of as place-based expressions of human culture.

In both natural and human-dominated systems, if these resilience behaviors are degraded, destroyed or were never present in the first place, the fragility of the system increases (it is more easily disrupted or damaged) because of reduced response, recovery and transformation capacity. As fragility increases, so does the need for interventions to sustain the system. The difficulty and cost of maintaining a fragile system tends to increase over time as the quality of internal resources declines. This is especially true when the system experiences more variable and extreme conditions, disruptions and damaging shocks. These patterns of resilience and fragility are currently playing out on a global stage as COVID disruptions and shocks reverberate through failing global networks including transportation, health, finance, energy, food and climate.

Resilience thinking provides some useful concepts and a new language that we can use to understand the underlying causes of the widespread and growing fragility of our lives and what must change in order to sustain the well-being of land, people and community in these unprecedented times. There are two key qualities that cultivate resilience in any kind of system: a wealth of diversity and a special pattern of relationships. These qualities support four system behaviors essential for the resilience of the system: self-organization, self-regulation, appropriate connectedness, and place-based innovation.[1]

These qualities and behaviors generate a wealthy and diverse portfolio of resources that support the well-being of the system over the long term. They can be observed, measured and managed in systems of all kinds, including agroecosystems.

The Raw Material of Resilience: Diversity

Although the importance of diversity to the well-being of agriculture and food systems is widely accepted, actually applying this principle is less well-developed. Resilience thinking can help us dig a little deeper into why and how a diversity of resources—natural, human, social, physical and financial—enhances resilience so that we can apply a more sophisticated understanding of how to cultivate resilience in food and farming systems (or any other social-ecological system). There are four different kinds of diversity that act together to sustain the well-being of a system in the face of disturbance: functional, response, spatial and temporal. These four kinds of diversity are observed in both natural and social-ecological systems.

Functional diversity describes the many different kinds of services produced by the species or communities of species (including humans) who engage in the agroecosystem in some way. Functional diversity in natural resources produces ecosystem services such as soil health, pollination and carbon sequestration.[2] Functional diversity can also be used to describe the different kinds of work that people do to keep the agroecosystem functioning properly—for example, crop planning, bookkeeping, field work, harvesting, processing and marketing. Farmers and ranchers can reduce production and marketing risks associated with disruptions of all kinds—not just weather—with some attention to cultivating high functional diversity across the full range of resources needed to maintain the well-being of the agroecosystem.

Response diversity describes the ability of the agroecosystem to maintain functional diversity over a wide range of environmental and social conditions. This is made possible by cultivating a diversity of species, or people or equipment that can produce needed services as conditions change. For example, the use of diverse mixes of cover crop species—called cover crop cocktails—reduces the risk of cover crop failure in the event of adverse soil or weather conditions. The mixes are designed with enough species diversity that some species in the mix will thrive no matter how soil conditions vary across the field or how variable weather is during the growing season. As another example, with the right training, the people working in the agroecosystem will be able to fill in for one another if needed.

The possibility of new federal food safety standards that would require organic growers to exclude all animals—wild or domesticated—from their production fields got Jim Koan thinking about new products that would allow him to continue to pasture pigs in his orchard at Almar Orchards and Cidery in Michigan. "We raise pigs because they're part of the system. I use them for insect and disease control, and I can sell their meat as a protein source. Almost all of our apples are processed on the farm and made into juice. Fifty percent of that bushel is still food. Even though the juice is taken out and fed to humans, you've got all this other good nutrition left in the pomace. That goes to feed our pigs, and then we use their manure for fertility in the orchard. We work as a team. It looks like with the new food safety regulations I won't be able to raise livestock on my farm anymore. For a sustainable farm, you have to have an integration of livestock and crops."

∽◌◌◌◌ ◌◌◌◌

Another way to think about response diversity is to think about it as investing in a backup plan or adding some redundancy that will help keep the agroecosystem going across a wide range of conditions. For example, a farm might enhance response diversity with a backup generator to provide power in the event the public grid fails, or by digging ponds as a backup water supply. Such investments in response and recovery capacity may decrease agroecosystem efficiency measured as simple annual net profit, but they are an important strategy for sustaining profitability despite variable conditions across the farm or from year to year. Continuing the cover crop cocktail example, it could be argued that the efficiency of the system was reduced by the decision to include species in the mix that ultimately did not germinate because of adverse soil or weather conditions. Likewise, the cost of the backup generator could be viewed as inefficient use of resources—because ideally it would rarely be needed.

Resilient systems tend to accumulate reserves across all types of capital. Like managing for diversity, investing in recovery reserves challenges traditional notions of economic efficiency because it redirects resources away from immediate production goals; however, recent research suggests that resilience and efficiency goals are complementary and a balance of both is required to enhance the resilience of the system[3] (see Figure 8.1). As climate change effects increase uncertainties, disruptions and shocks, investments in response and recovery capacity are rapidly becoming part of best practice in business.[4]

Spatial and temporal diversity enhance functional and response diversity by providing opportunities for the formation of diverse ecological and social relationships across space and time. Agroecosystems with high ecological spatial and temporal diversity have a number of distinctly different small-scale ecosystems present in production fields, pastures, edges and natural areas that change over time. This landscape mosaic features patches of both intensively managed and unmanaged land integrated into a whole through the movements of livestock and wildlife. An example of high social spatial and temporal diversity would be the

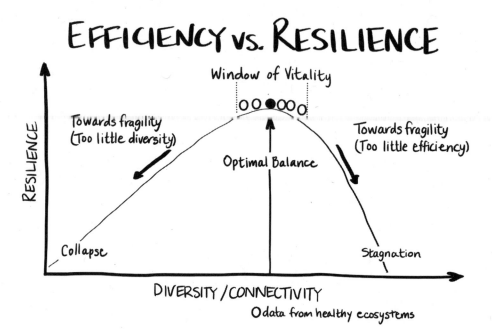

EFFICIENCY vs. RESILIENCE

Window of Vitality

RESILIENCE

Towards fragility
(Too little diversity)

Optimal Balance

Towards fragility
(Too little efficiency)

Collapse

Stagnation

DIVERSITY/CONNECTIVITY

O data from healthy ecosystems

Figure 8.1. Healthy natural ecosystems develop networks of energy, material and information flows that have sufficient diversity to create the functional and response diversity needed to maintain ecosystem identity in changing conditions without wasting resources on unnecessary redundancy. Natural ecosystems that display this optimal balance of diversity and efficiency are more resilient than ecosystems that are more efficient or more diverse.[5] Agroecosystems could be designed and managed to mimic the flow patterns found in the window of vitality. Credit: Caryn Hanna.

presence of multiple generations living on the farm, each responsible for managing different core enterprises. It is the people in the agroecosystem that integrate individual enterprises into a whole that persists over the long term.

"A testimony to the resilience of our farm system is the fact that we have never lost a crop to pests in 30 years now of no pesticides. We have never even come close to having insects or disease or anything destroy our yields. We've had stable, very good yields during all that time. That's always something that people tend to look at you with, 'Huh? I find that hard to believe.' People don't believe it. Of course, I'll be the first to say we have a lot of things we could be doing better, but still, overall, the resilience shows itself in our system. Our soil quality shines here. It really does. And we work to enhance those ecosystem services, as they're called, by continually planting more trees, more shrubs, more crops for pollinators, more windbreaks and more wildlife habitat. The diversity is what will continue to play a big role for us."

—Ron Rosmann, Rosmann Family Farms

A Diversified Portfolio of Assets

System resilience is associated with the amount, diversity and quality of resources or capital assets—natural, human, social, financial and physical[6]—available for use to cope with, adapt to or transform a system in response to stress, disturbance and shocks.[7] Access to an abundance of high-quality resources across the full range of asset types offers the greatest opportunity for sustaining the system under variable and changing conditions. A healthy natural resource base provides climate protection services and enhances the response diversity of the system. High-quality human and social resources enhance the learning and innovation capacity required for innovative responses to challenging conditions. Sufficient financial and physical resources provide access to the necessary labor, management, tools, equipment and technologies to put innovative solutions into action.

RESILIENT

Natural   Human   Social   Financial   Built

Key Resilience Behaviors

Farmers and ranchers can get quite a way along the path to resilience just by focusing on the design and management of operations that cultivate high functional and response diversity and make it possible to accumulate a wealthy portfolio. But it may be helpful for agricultural producers and the people who support them to dig just a little deeper into why diversity and wealth are so important to resilience. It turns out that these two key qualities of resilience support four system behaviors that are critical to maintaining the identity[8] of the agroecosystem in challenging conditions. These behaviors are self-organization, self-regulation, appropriate-connectedness and place-based innovation.

Self-Organization

Self-organization describes the spontaneous emergence of a recognizable order created by the self-interested actions of individual organisms in a system. Each independent organism in the ecosystem interacts with others to obtain needed resources, and out of these relationships recognizable patterns begin to emerge. Farmers and ranchers benefit from ecological self-organization when they inoculate legume seed with rhizobia before planting, or use managed grazing to restore grasslands. Many sustainable agriculture practices cultivate the self-organizing behaviors of agroecosystems to produce the resources needed for the healthy growth and development of crops and livestock.

Self-organization can be observed across many scales—from micro to planetary—and in natural, social-ecological and social systems. For example, self-organization cultivated the characteristic patterns of the Earth's ecosystems (see Figure 8.2) as well as the Indigenous foodways[9] associated with each. Self-organization is responsible for the diversity of human cultures, each with their own distinctive qualities. Some sustainable farmers are using these social-ecological patterns to restore natural landscapes degraded by industrial agriculture, to guide the design and management of their farming systems, and to develop new crop and livestock cultivars that are better suited to local ecological conditions.

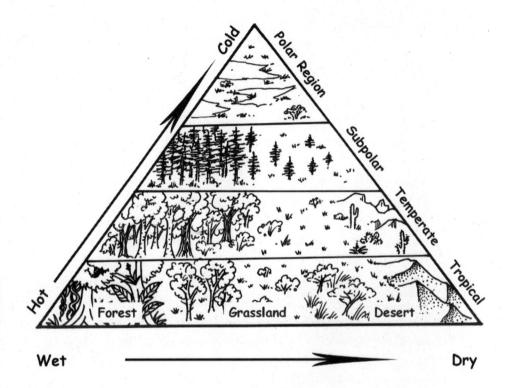

Figure 8.2. The land-based ecosystems of the Earth can be grouped into four major biomes based on dominant vegetation—tundra, forest, grasslands and desert—that vary by temperature and precipitation. Pastoralism evolved in grasslands, shifting horticulture evolved in forests and sedentary agriculture evolved in both temperate and desert regions with access to manageable water supplies. Credit: Sherri Amsel.

Self-Regulation

Feedbacks are relationships that govern the behavior of a system through interactions between the components or parts of the system that alter ecological and social processes. These interactions typically involve exchanges of information, energy or materials between two or more components. Feedbacks in ecological and social systems are typically balancing forces[10] that operate to maintain the system in its current state. Temperature regulation is a classic biological example of a balancing feedback. If your body temperature moves outside of a normal range, temperature regulation systems send governing signals to activate responses that decrease (sweating) or increase (shivering) system temperature. Balancing forces can also occur in social systems, for example when citizens send signals to politicians with their votes, and in physical systems, for example, a home heating system.

Much rarer in systems are self-reinforcing feedbacks, which act to push the system out of balance and to reinforce a change in the system.[11] The longer the reinforcing feedback is active, the greater its capacity to create fundamental change. Reinforcing feedbacks are powerful forces of change that drive growth, explosion, erosion and collapse in ecological and social systems.[12] Population growth is the classic example of a biological self-reinforcing feedback: the more babies born, the more people grow up to have babies. Disease outbreaks are another: the more people infected with COVID, the more people there are to spread the disease to other people. Soil erosion is also self-reinforcing: as erosion reduces soil health, erosion processes increase in intensity leading to a downward spiral of soil degradation.[13] Self-reinforcing feedbacks also occur in social systems, for example the interest earned on a savings account, the erosion of the middle class or the tendency for the wealthy to get wealthier.[14] Left unchecked by balancing forces, self-reinforcing feedbacks ultimately transform the existing system into a new one with a different structural and functional properties.

Social-ecological systems typically have numerous balancing feedbacks acting to maintain the identity of the system as conditions change. Some of the balancing feedbacks are inactive most of the time—you could compare this to the sweating and shivering feedbacks that maintain your body temperature—but are crucial to the system's well-being (in this example, your body). Agroecosystems with a high degree of ecological self-regulation can maintain agroecosystem performance over a wide range of conditions without the need for additional management, energy or material subsidies.

As social-ecological systems develop, there is often a trend toward weakening balancing feedbacks by eliminating them (in the quest for efficiency) or by increasing the distance the feedback must travel as energy and material use and governance become more complex and increase in scale.[15] Management strategies that cultivate balancing feedbacks do so at the expense of industrial notions of efficiency, because resources must be invested in components that are inactive much of the time.[16] Systems that emphasize efficiency over balancing feedbacks and strive to protect self-reinforcing feedbacks—especially those that promote the inequitable accumulation of wealth—face a growing risk of catastrophic failure in a changing climate.[17]

Resilient Patterns of Connection

Resilience thinking invites a focus on the number and quality of ecological and social relationships within the agroecosystem and between the agroecosystem and other systems. Resilient systems cultivate a specific pattern of internal and external relationships that enhance functional and response diversity. In resilience thinking this pattern is known as modularity. Another name for resilient networks of relationship is "appropriate-connectedness," to reinforce the idea that relationships that enhance functional and response diversity are "appropriate" if the goal is resilience.

Modularity describes two kinds of connections between the parts that make up a system and between that system and other systems. Tight connections are relationships that transmit information, energy and materials critical to the health of the system. Loose connections are relationships that may benefit one or both parts, but are not critical to the well-being of either.

At a particular scale, appropriately connected systems have tight connections within each component of the whole system, but those system components are only loosely connected to each other. This arrangement of system components increases the independence of the components and enhances the capacity of the whole system to function despite disturbances to one or more components. Modularity also encourages response diversity because each component in the system, being only loosely connected to the rest, is free to self-organize and innovate unique processes and operational strategies.

For example, in the early months of the COVID pandemic when food service businesses abruptly shut down, industrial producers and processors that were tightly connected to national networks of food service buyers had no choice but to dispose of massive amounts of product.[10] The nature of their connection to the industrial food system gave them no flexibility to pivot to a different market. In contrast, sustainable producers with loose connections to multiple buyers were able to adjust to the new market conditions more easily—often aided by local and regional food organizations that swiftly created new online marketing options and innovative distribution networks such as drive-through farmers markets. As a result, many growers serving local and regional direct markets reported record sales in 2020.[19]

From a resilience thinking perspective, the use of sustainable practices makes it more likely that a farm or ranch will cultivate tight relationships on the farm and loose relationships with suppliers and markets. This is because sustainable practices tend to cultivate functional and response diversity. Because sustainable producers are invested in creating ecological relationships that produce beneficial ecosystem services, sustainable farms tend to be relatively self-sufficient when it comes to production inputs. Because sustainable producers use diversified marketing practices and prefer high-value direct and wholesale markets, sustainable farms tend to be less reliant on any one market, but instead cultivate loose relationships with a number of markets.

In contrast, industrial thinking encourages the use of standardized components that are tightly connected in a global network that depends on an uninterrupted global flow of information, energy and materials. Standardization increases the fragility of the system, because a disturbance or shock can be swiftly transmitted throughout the whole system and deep into all components of the system through cascading component failures. For example, a farm that assumes optimum growing conditions and substitutes standardized technology for ecosystem services—for example, a technology package of genetically modified corn cultivar, fertilizer, pesticide and irrigation as a substitute for functional and response diversity—is at high risk of total crop failure if the supply of just one of the many globally-sourced inputs critical to the package is disrupted. This strategy of corn production is so fragile (and growing more so every year) that producers using industrial production strategies must shift the risk of crop failure to the public through the purchase of government-subsidized crop insurance in order to stay in business.[20]

Tom Trantham appreciates the flexibility offered by the highly diverse, pasture-based system that he employs on his pasture-based dairy in South Carolina. Trantham invented a pasture-cropping system he calls "12 Aprils" that features no-till planting of seven or eight short-season annual forages into a perennial grass pasture each year. "I prepare for what I think the situation's going to be," Tom says, "and then if it doesn't work, I just bushhog it and plant something else. That's the great thing about my system." The ability to recover quickly from mistakes or the unexpected has been particularly helpful over the years. Using no-till also provides a lot of flexibility, plus it saves time and money in fuel and equipment costs. "There's always a challenge in farming," he says, "but if you make a mistake…or maybe it isn't a mistake, maybe it rained too much or it was too dry, with my system you're not set back too much. Just the number of days it takes for you to get back out there and replant. But when you've got a hundred acres of corn silage, and you lose it, you don't have another shot until next year, so you're done for. You've got to buy feed and all, and that'll break you in a heartbeat, to have to purchase feed."

࿐࿐࿐࿐ ࿐࿐࿐

Large-scale, tightly connected systems are vulnerable to spectacular, system-wide failures like the East Coast blackout of 2003,[21] the global financial meltdown of 2008,[22] the late blight pandemic in the Eastern United States in 2009,[23] and the many product, service, market and economic failures associated with the COVID pandemic in 2020.[24] Resilience thinking reminds us such failures are more likely as these systems grow larger and more complex.[25,26] In addition, climate change is likely to increase the frequency and intensity of system-wide failures.

Designing and managing agroecosystems to cultivate appropriate-connectedness has a very challenging implication: farmers and ranchers must find ways to achieve their management goals largely through local and regional social-ecological services. Managed well, agroecosystems that make use of appropriate patterns of connection are more likely to restore, conserve and accumulate community-based wealth and to avoid contributing to the exploitation and degradation of other regions through the import of needed resources or export of wastes. Practices that promote high functional diversity—for example, the use of cover crops to produce the nutrients needed by the agroecosystem—also enhance soil health on the farm, improve regional water quality and can reduce or eliminate the need to import nutrients from faraway places. An agroecosystem that cultivates many tight social connections at local and regional scales and a few loose connections at the national and global scales will be less vulnerable to disturbances and shocks caused by social conditions that are outside the influence of the producer.

Cultivating Place-Based Innovation

The final key resilience behavior is the ability to learn as you go. This is a special kind of learning by doing that uses careful observation and record-keeping to understand how place shapes the response of the agroecosystem to changing conditions coming from inside and outside the system. All living systems have the capacity to learn from disturbances. The careful observation of agroecosystem responses to frequent, small-scale disturbances will stimulate and strengthen ecological response and recovery processes.

Agricultural managers can take the opportunity to observe and learn from agroecosystem responses to disturbances, disruptions and shocks in order to select robust plants, animals, production and marketing practices and operations designs that cultivate high response, recovery and transformation capacity. For example, the practice of integrated pest management helps producers improve natural pest suppression in their agroecosystem through a learn-as-you-go process that tolerates low pest populations to promote ecological and social self-organization and self-regulation.

Reflective learning is a special kind of community-based learning that cultivates the resilience of agroecosystems. Reflective learning encourages the blending of scientific knowledge with different kinds of local knowledge—Indigenous, colonial, traditional and contemporary—to create place-based innovations that are well-adapted to the land, people and communities living in and dependent on the agroecosystem.

 Thinking about the future, New Mexico rancher Julia Davis Stafford feels fairly confident in the management practices she uses to reduce the risks of weather variability and extremes, particularly planned grazing, soil health, water conservation and the use of drought-tolerant forage varieties and cattle that are well adapted to the region. Julia says that if climate change continues to intensify, she'll likely just continue to destock the ranch, figure out how to cut back on the need for irrigation and how to supply water to the remaining stock if surface waters were to fail.

Julia also plans to keep learning how to improve existing management practices and about new practices through participation in groups like the Quivira Coalition. "What is always tremendously encouraging to me is just the networking at these various agricultural gatherings, talking to people, and going to listen to them speak," Julia explains. "Sometimes, particularly just after I get home from a Quivira Coalition conference, I feel we'll be able to sort through this and go on just fine."

ᖇᎧᎧᖇ ᎧᖇᎧ

Reflective learning also recognizes the powerful influence that past conditions and experiences have on the present identity and future evolution of the agroecosystem. The biological and cultural memory embodied in an agroecosystem is a rich source of place-based knowledge and experience that can be drawn on for inspiration and innovation. This aspect of reflective learning is particularly important in current efforts to acknowledge and repair the harms of industrialism and embrace the wisdom of Indigenous cultures.

 The longer growing season and warming temperatures at the Acorn Community's farm in central Virginia have enhanced Ira Wallace's success working with tropical perennials for the Southern Exposure Seed Exchange. "A lot of our tropical perennials have been brought by people from all over the world, by different ethnic groups," says Ira. "I'm particularly interested in things from the African diaspora. This work is not just about promoting traditional plants, it is also about retelling the story of agriculture, starting with African people being brought here because of their experience as farmers and not just as brute labor in the Southeast." Southern Exposure is currently leading a national project to recover and share traditional collard varieties.

∽◌◌∽ ◌◌◌

Practices that support reflective and shared learning in agroecosystems include adaptive management strategies, the preservation of Indigenous and traditional knowledge, and participation in producer-based research, education and marketing networks. The maintenance of local heirloom crops and livestock breeds, engagement with elders and the exploration of Indigenous and traditional foodways are just a few examples of activities that seek to acknowledge and incorporate the cultural and ecological heritage of place in order to enhance the resilience of agroecosystems.

 Southern California vegetable grower A.G. Kawamura says that many years of growing as part of the growers' network that supplies OC Produce has increased the resilience of his operation. "There's nothing that can replace experience and observational knowledge of a region," says A.G. "Actually knowing your area and being able to recognize vulnerabilities or opportunities. We share and exchange information across our companies and we work collaboratively to help each other. As a good example, we had a disease complex I couldn't identify. I texted a picture of it to our partner down in Mexico and he immediately ID'd it as a new mildew and gave us a whole new prescription on how to handle it. It was a perfect example of how quickly I can find some help when I'm dealing with something I haven't seen before. We must keep expanding our knowledge base."

Years of active research and development work on his own farm in north central Montana has left organic grain producer Bob Quinn a little frustrated at the slow erosion of support for government technical assistance programs like the Natural Resources Conservation Service and the Cooperative Extension Service. "I wish that we could send that message to Washington somehow, to tell them that they don't have all the answers, and by pretending that they do, they really stifle imagination and response to some of these future challenges. Especially now, when things are obviously changing, we need to really be thinking of as many different solutions as possible, not trying to apply a solution that worked in the last decade. Many of those old solutions were fine, but they might not work in the future nearly as well as something that some people are only just starting to think about now, or something that hasn't even been thought of yet."

Specified and General Resilience

There are two kinds of resilience that can be managed in agroecosystems and both are needed to reduce vulnerability to changing conditions. *Specified resilience* reduces the threat of a specific risk to a specific component of the agroecosystem, for example, the resilience of the spring bloom to a late frost, the resilience of a bare soil to erosion from heavy rains or the resilience of pastures to drought in summer. *General resilience* is the capacity of a system to respond to disturbances of all kinds, including those that are unexpected or unmeasurable, in a way that maintains its existing structure, function and purpose.

Specified resilience is the kind of resilience people are thinking about when they consider the question: the resilience of what, to what? Cultivating specified resilience involves weighing the likelihood of a specific kind of disruption, the likely impact of that disruption on operations and the cost associated with preventing or moderating the damages associated with the disruption. For example, the costs of constructing and maintaining a levee to protect a farm from losses due to flooding can be accurately estimated, as can the likely losses from flooding (although the risks associated with such events are increasingly difficult to project based on historical data).

Managing for specific resilience is very similar to traditional agricultural risk management because of the focus on moderating specific, high-impact threats to the agroecosystem. It differs by encouraging producers to consider the costs and benefits of the full range of resources under management—natural, human, social, financial and built/technological—rather than relying on a short list of favored risk management tools that may have worked well in the past. Managing for specified resilience is an appropriate risk management strategy when the likelihood, costs and benefits associated with a specific kind of disturbance can be reliably estimated; however, too much focused investment on a specific threat tends to degrade the general resilience of the agroecosystem.[27] After identifying and managing for specified resilience to high likelihood/high impact threats, resilience thinking turns to practices that enhance general resilience.

General resilience is what most people think of when they refer to resilient people, agriculture or landscapes. General resilience is the capacity of a system to respond and recover from disturbances of all kinds, including unforeseen and unprecedented events, in order to preserve its identity. Rather than reducing the risk of damage from a specific kind of threat, managing for general resilience strives to enhance the overall coping capacity of the system. It is well aligned with the concept of adaptive capacity as defined in vulnerability assessment.[28]

Resilience Design Principles

As climate change effects become more damaging to society, there is a growing interest in applying resilience thinking to the assessment, design and management of many kinds of systems. Businesses are working to secure resilient supply networks,[29] cities are exploring resilient improvements in infrastructure,[30] landscape architects are designing resilient landscapes,[31] and governments at all levels—from international to local—are expanding climate change mitigation and adaptation planning to include resilience objectives.[32] This important work has expanded our understanding of the qualities and behaviors of resilient systems by exploring the practical application of resilience principles to design and management practice across a diversity of settings and has generated some useful new tools for the assessment, design and management of agroecosystems.[33]

For example, Table 8.1 presents a set of agroecosystem qualities and behaviors, expressed as resilient design criteria, that I have found particularly useful as a tool for digging into how sustainable agricultural practices might promote food and farming resilience. The rest of the table includes some specific examples of sustainable agriculture practices and sustainability indicators that support each criteria.

A careful review of Table 8.1 helped me understand how sustainability and resilience are related and how they are different. As I gained experience applying these resilience criteria to the design and assessment of agroecosystems, the difference between sustainability and resilience thinking became clear to me. Resilience thinking goes deeper into the sustainability principle of diversity to include a more explicit focus on the importance of functional and response diversity to the response capacity of agroecosystems. In addition, resilience thinking invites more attention to cultivating specific patterns of networks of relationship (appropriately connected) and recognizes the capacity of an agroecosystem to learn from disturbance. Simply put, resilience thinking takes us beyond sustainability in exactly the ways that we need to sustain agriculture in a changing climate.

Table 8.1. Resilience Design Criteria for Agroecosystems

This table shares an indicator framework for the design and assessment of agroecosystem resilience as set of resilient qualities/behaviors expressed as design criteria and some examples of common sustainable agriculture practices and farm sustainability indicators (measures of agroecosystem performance) associated with each criteria.[34]

Resilience Design Criteria	Associated Sustainable Agriculture Practices	Prosperity Project Sustainability Indicators
Ecologically self-regulated	Farm maintains diverse annual plant cover and incorporates perennials, provides habitat for beneficial organisms and aligns production with local ecological conditions.	Soil Quality, Balanced Nutrient and Carbon Budget, Energy and Water Efficiency, Pest Pressure
Functional and response diversity	Diverse crop rotations, integrated and pasture-based livestock production systems, composting, alternative energy production, water harvesting.	Soil Quality, Balanced Nutrient and Carbon Budget, Energy and Water Efficiency, Pest Pressure
Spatial and temporal diversity	Farm landscape is a mosaic pattern of managed and unmanaged land, diverse plant types and livestock are cultivated across space and time, diverse crop rotations integrated with livestock.	Biodiversity
Appropriately connected	Collaborating with multiple producers, suppliers, markets and farmers; farm design that encourages response diversity.	No comparable indicators
Exposed to disturbance	Management that accepts some controlled disturbance from weather variability, nutrient variability and pests in order to discover robust crops, livestock and production system configurations.	No comparable indicators
Coupled with local natural capital	Farm builds soil quality to maintain healthy water and mineral cycles, has little need to import water or nutrients or export waste.	Soil Quality, Balanced Nutrient and Carbon Budget, Energy and Water Efficiency
Socially self-organized	Farmers and consumers are able to organize into grassroots networks and institutions such as co-ops, farmer's markets, community sustainability associations, community gardens and advisory networks.	Participation and Cooperation in Community
Builds human capital	Investment in infrastructure and institutions to support community-based education, research and development and local businesses; support for social events in farming communities.	Time for Family Activities, Family Education, Farm Succession Plan, Local Sales, On-Farm Jobs, Local Purchases, Community Cooperation, Local Identity

Table 8.1. (cont'd.) Resilience Design Criteria for Agroecosystems

Resilience Design Criteria	Associated Sustainable Agriculture Practices	Prosperity Project Sustainability Indicators
Reflective and shared learning	Extension and advisory services for farmers; collaboration between universities, research centers and farmers; cooperation and knowledge sharing between farmers; record keeping; baseline knowledge about the state of the agroecosystem.	Cooperation with Other Farmers, Community on Farm, Family Education
Honors legacy	Maintenance of heirloom seeds and engagement of elders, incorporation of traditional cultivation techniques with modern knowledge.	Local Identity
Globally autonomous and locally interdependent	Less reliance on commodity markets and reduced external inputs; more sales to local markets, reliance on local resources; existence of farmer co-ops, close relationships between producer and consumer, and shared resources such as equipment.	Local Sales, Local Purchases, Community Cooperation, Community on Farm
Reasonably profitable	The people involved in agriculture earn a living wage, a reasonable return on invested capital, have the resources needed for healthcare, education, family activities and retirement; farm work brings a feeling of satisfaction to the people working on the farm.	Total Family Income, Time for Family Activities, Family Health, Satisfaction from Farming, Farm Succession Plan

9

The Rules of Resilience

As I learned more about the ecological and social behaviors of resilient systems and started using them in my work, I slowly began to see that the key qualities and behaviors of resilient systems—diversity, community wealth, self-organization, self-regulation, modularity and innovation—could be summed up neatly in a few rules describing three behaviors that promote the resilience of social-ecological systems:

- Cultivate diverse networks of reciprocal relationship
- Cultivate regional self-reliance
- Cultivate the accumulation of community-based wealth

I have found power in these rules. They have helped me navigate the noisy and confusing community that has sprung up in the last decade around food system resilience, agricultural climate solutions and community resilience. These three rules have both challenged and changed my thinking about some of my own favored climate solutions by inviting me to dig deeper into the fragility of industrialism.

Be warned: these rules are not easy to live by! Following these rules will most likely get you into some "good trouble, necessary trouble" to quote John Lewis.[1] I know. I've been doing my best to use them in my own life—at home, in my business, in my work and in my community.

These rules are not kind to sloppy thinking or the kind of thinking that fails to keep the whole in mind. They are not gentle with thinking that refuses to acknowledge our dependence on nature and the very real suffering that makes our good life possible. These rules will require you to take an honest look at the underlying cultural values that have set our world on fire. They are not easy to live by, but if embraced with trust, kindness and courage, these three rules of resilience have the power to heal the wounds of industrialism and sustain the well-being of land, people and community through the inevitable challenges that lie ahead.

Diverse Networks of Reciprocal Relationship

The first rule of resilience invites us to remember that resilient systems are made up of diverse networks of mutually beneficial relationships that exchange information, energy and materials. These relationships span the local to the global and include all our relations: the land, plants and animals (including humans) within our own community and the relationships between our own community and other communities. These relationships span all time, stretching back to the people that lived before us and forward to those that have not yet lived.

Expressions of this first rule of resilience, reminding us about the necessity of healthy relationship in community, are found in many cultures, across all time:[2]

- "I am because we are."
- "Do unto others as you would have them do unto you."
- "Healthy soils, healthy plants, healthy people, healthy communities, healthy planet."

It will come as no surprise that this rule is rooted in the kinds of relationships that cultivate the well-being of nature—diversity, reciprocity, health, abundance—because we are part of nature. We learned this truth a long time ago, back in a time when we lived within the ecosystems that we called home. Back in the days when we understood that our own lives depended on the gifts shared with us by land, people and community.

Because we are human, our way to remember and live by this truth is to name and encourage the kinds of behaviors that cultivate healthy relationship in community. You can think of these behaviors as the core

values of healthy community life. These values seem to be universal—
they are found in all cultures, across all time and in every part of the
world.[3] It seems that at a very basic level, although expressed with slight
differences, we all agree with this common ethical code:[4]

- Honesty
- Equity
- Responsibility

- Compassion
- Respect
- Courage

These six behaviors cultivate the kind of personal relationships that are
essential to community resilience. For example, trustworthy people are
not only honest and equitable in their relationships with others, but they
also acknowledge and accept their share of responsibility for the well-
being of the whole community. Compassion and respect extend the practice of trust by cultivating kindness. Kindness requires empathy—the
willingness and ability to put yourself in the place of another and then to
treat that person as you wish to be treated.

It takes courage to be trusting and kind, especially in a culture so
wounded by the myth of individualism,[5] so enchanted with ridicule and
so willing to celebrate winning at any cost. Acting with courage extends
the practice of kindness by cultivating self-confidence, discipline and
perseverance. It takes courage to form meaningful relationships with
other people and to stay committed to those relationships through times
of inevitable misunderstanding, disappointment and hurt feelings. It
takes courage to be brave.

Although I've discussed these essential values in terms of human relationships, they apply equally to our relationships with all of nature. If
you are not sure how to get started rethinking your relationships, I encourage you to dive into the very wise and constantly evolving conversation about a just transformation of food and farming currently underway
among diverse communities of food producers, activists, organizers and
scholars in the U.S. and beyond.[6] In recent years, new perspectives on
food justice,[7] the integration of traditional and scientific knowledge[8] and

the decolonization of industrial culture[9] have been particularly helpful to me as I've explored how food and farming will have to change in order for us to abide by the first rule of resilience.

 Vegetable grower C. Bernard Obie, called "Obie" by his friends, views his relationships with his crops as the most important resilience asset on his farm in north central North Carolina. "We try to prepare plants for what it is that we're calling them into the world for," says Obie. "We engage them to help us with our mission and we promise them we will do our best as well." Although Obie recognizes that some biodynamic practices, like the Stella Natura calendar, are "not scientific," his experience confirms that working according to the calendar is helpful. A second practice that Obie uses to cultivate resilience on the farm is "just taking the time when you're out in the field to observe, to be quiet and just be in the space to get a feel for how things are going."

Regional Self-Reliance

The second rule of resilience invites us to remember that resilient systems must be able to sustain our community without relying on the import or export of energy, materials or information. This rule encourages respect for the social-ecological limits of place in the design and management of community systems that are critical to our well-being. Practically speaking, the second rule of resilience invites us to imagine how we can live well within the resources available in our own region without the need to exploit land, people and community in other regions. Notice that this rule is also rooted in the kinds of relationships that promote the well-being of nature.

Cultivating regional self-reliance requires us to respect the ability of regional ecosystems to support healthy community. You can think of this ability as the place-based biocapacity of land and people.[10] The social-ecological biocapacity of a region represents the productivity of its ecological assets (including cropland, grazing land, forest land, fishing grounds and built-up land), plus the capacity of those assets to absorb wastes and to regenerate a new supply of assets. We can explore the biocapacity demand and supply of a community by measuring its ecological footprint.[11]

As a nation, our appetite for resources is about two and a half times our national biocapacity and has been for more than 50 years.[12] Because we run a biocapacity deficit, industrial thinking gives us no choice but to liquidate our own resources (using resources faster than they can be regenerated), import resources from other places and release wastes into the environment. Purchasing the biocapacity we need from the global industrial supply may seem like a good solution, until you realize that the rest of the world has also been running a biocapacity deficit for nearly as long as we have.[13]

This second rule of resilience—regional self-reliance—invites us to step back and reconsider the wisdom of industrial thinking that has, quite literally, pushed our planet beyond safe-operating limits.[14] It challenges us to imagine a different way to organize human communities, a way that respects regional biocapacity and restores the regenerative power of regional ecosystems. Changing our thinking about biocapacity in this way may seem like an impossible challenge, but recent research suggests that most U.S. metropolitan areas could feed themselves with a shift to healthier eating and the sharing of some foods between regions.[15] No matter how impossible living by the second rule of resilience may seem, this may soon be more than an academic question. If the Southwest and Southern Great Plains run low on water as expected, we may have no choice but to move food production to more water-wealthy regions of our nation.

Accumulation of Community-Based Wealth

The third rule of resilience invites us to remember that community-based wealth provides the resources that are needed to sustain community well-being over the long term. This rule directs our attention to the amount, diversity and quality of resources—natural, human, social, financial and physical—held within a community to cope with, adapt to or innovate in response to stress, disturbance, shocks or new opportunities.[16] A healthy natural resource base enhances the functional and response diversity of the community. High-quality human and social resources support the innovation capacity required to sustain community well-being in challenging conditions. Financial and physical resources provide the necessary tools, equipment and technologies to put innovative solutions into action and to recover from damaging events.

It should be obvious by now that the third rule of resilience requires completely rethinking the meaning of wealth. "Community-based" begins to express this difference of wealth in a broader sense—wealth that goes beyond financial assets to include all of the resources that promote the well-being of community. It turns out that this is a very old idea. My ancestors expressed this idea nearly a thousand years ago as "the commonweal," a Middle English word that invites us to remember that "wealth" originally meant "well-being," and that well-being is centered in community. The third rule reminds us that resilience requires attention to the commonweal, which means "the public good, the general welfare of the nation or community." Commonweal eventually evolved into the word commonwealth to describe an alliance of communities working for the general welfare of the whole.

The third rule of resilience demands that we find new ways to think, talk about and measure community-based wealth. Over the last 50 years, organizations promoting community sustainability have created many useful resources to support community design and management that are well-aligned with the third rule,[17] including resources developed to promote community-based sustainable food and farming.[18] More recently regenerative economics, a new kind of economics based on the flow patterns cultivated by healthy ecological relationships, offers a completely new way to think about community well-being.[19]

Three hundred years of industrial exploitation of land and people have left many communities throughout our nation facing growing climate risks with diminished resources. Climate resilience planners have recently begun to acknowledge the unique resilience challenges of these so-called frontline communities—historically underserved communities of Indigenous, Black, minority and rural people[20]—and have responded with assessments and planning efforts that typically include targeted economic development incentives, reparation initiatives and locally-led resilience assessment, planning and implementation.[21] Although the motivation for these new investments is to enhance the resilience of the commonweal, they also represent an example of the kind of unique opportunities for healing the wounds of industrialism that we will find on the path to a resilient future.

Figure 9.1. The three rules of resilience sum up the essential ingredients of resilient community. These rules are not easy to live by, but if embraced with honesty, trust, kindness and courage, these rules have the power to put us on the path to a just transformation. Credit: Caryn Hanna.

Moving Beyond Industrialism: A Just Transformation

There is growing agreement across a broad group of organizations, businesses and governments worldwide on the need for a fundamental transformation of food and farming in these times.[22] Although significant disagreement about preferred paths to this transformation remains, these discussions seem to be converging on a shared vision of food and farming that is sustainable, nutritious, inclusive and resource efficient. There is also agreement on who must be involved to achieve this transformation: government, business, civil society and consumers—in other words, all of us! Governments must shift public investment, business must shift operational strategies, civil society must shift mission and consumers must shift what we eat.

As a specific example, the UN Food and Agriculture Program regularly convenes diverse stakeholder groups and invites them to find areas of consensus on both problems and solutions in the global industrial food system.[23] The goal of the program is not to broker compromise among these groups, but rather to find new ground from which to move forward together on innovative solutions. In 2012, stakeholders representing Agribusiness, Rural Livelihoods, Environmental Sustainability and International Trade agreed that for the first time at the global level agriculture faces multiple natural resource limits: land availability, water quality and quantity, soil quality, agrobiodiversity, energy and a stable climate. Although there were some areas of disagreement on the future vision, these stakeholders found new ground for collaboration around these transformative goals:[24]

- Significantly reduce food waste
- Redefine the purpose of agriculture as human nutrition
- Support small- and medium-sized farms
- Manage for high yields in healthy ecosystems
- Develop appropriate technologies for a diversity of contexts
- Reward business performance that enhances the public good
- Measure whole system sustainability
- Adapt institutions to support these goals

This universal recognition of the need for food system transformation has followed on the heels of a slow but steady global awakening to the unprecedented challenges of climate change. This global discussion about how we change the way we eat is both welcome and long overdue; however, it is important to remember that nearly everyone engaged in these conversations is limited by sustainability thinking that puts more or less of a premium on maintaining "business as usual." From a resilience thinking perspective, most of these discussions tend to generate solutions designed to adapt the global industrial food system to climate change rather than more transformative solutions.[25]

Recall from the first chapter those six major shifts that will move us beyond industrialism. They are listed again below according to the level of challenge they represent to industrial thinking, from low to high, along with a brief discussion about why the shift is a challenge and some specific examples of agricultural practices that illustrate the shift. This discussion amplifies the differences between industrial and resilience thinking in order to highlight some distinctive behaviors of each. In practice, both kinds of thinking draw from the same toolbox and sometimes use the same tools to achieve similar goals. How do your favored agricultural climate solutions promote the shift to resilient agriculture?

- **From optimum to variable conditions**—Industrial thinking strives to control variability with technology in order to achieve ideal production conditions. Resilience thinking strives to enhance the capacity of the agroecosystem to moderate, tolerate, and benefit from variability. For example, industrial thinking responds to declining water resources by encouraging a shift to more drought-resistant crop cultivars, water conditioning and precision irrigation equipment, and moving operations into physically-protected, climate-controlled structures. In contrast, resilience thinking encourages a shift to production system designs that eliminate the need for irrigation and promote the health of the regional water cycle.
- **From efficient to redundant systems**—Industrial thinking strives to maximize efficiency through economies of scale and to transfer

the costs of operational fragility to others. Resilience thinking strives to maximize resilience through economies of scope[26] and to manage the costs of operational fragility internally. For example, industrial thinking responds to growing costs of weather-related disruptions by increased participation in publicly-subsidized agricultural insurance and recovery programs. In contrast, resilience thinking encourages a shift to more diversified operations that reduce the costs of weather-related disruptions by reducing agroecosystem sensitivities to weather-related disruptions.

- **From industrial to ecological logic**—Industrial thinking strives to avoid the uncertainties associated with relying on healthy ecosystem services to produce marketable goods. Resilience thinking strives to take advantage of healthy ecosystems services in order to produce marketable goods. For example, industrial thinking encourages the substitution of technologies like fertilizers and pesticides to meet the nutritional and pest suppression needs of agricultural operations. In contrast, resilience thinking encourages cultivating ecosystem services to meet these needs through the use of practices like diversified crop rotations, cover cropping and the integration of crops and livestock.

- **From expert to place-based knowledge**—Industrial thinking strives to support uniform operations regardless of local conditions. Resilience thinking strives to position operations to take advantage of the unique qualities of place. For example, to reduce the risk of drought, industrial thinking encourages the use of improved, drought-resistant crop cultivars well-suited to industrial operations. In contrast, resilience thinking encourages the use of drought-tolerant crop species and livestock breeds that are well-adapted to local conditions and well-suited to diversified operations.

- **From global to regional resources**—Industrial thinking strives to maximize financial return by managing low-cost, globally-sourced resources to produce high volumes of uniform products that meet global market demands. Resilience thinking strives to maximize returns on a diversified portfolio of regional resources—natural, human, social, financial and built—by managing those resources to produce unique place-based products that meet regional market demands. For example, industrial thinking encourages the production of commodity corn and soybean feed grains for global markets with imported resources. In contrast, resilience thinking encourages the production of diverse species of regionally-adapted cultivars of food and feed grains to meet regional grain needs with regional resources.

- **From extractive to regenerative economy**—Industrial thinking strives to deliver high returns on operations through practices that rely on the liquidation of community-based wealth. Resilience thinking strives to deliver high returns on operations through practices that generate community-based wealth. For example, industrial thinking encourages increasing returns by employing temporary migrant labor at less than a living wage, minimizing the costs of meeting health and environmental regulations, and taking advantage of public subsidy and assistance programs. In contrast, resilience thinking encourages increasing returns by employing permanent full-time labor at a living wage, investing in health and environmental practices that go beyond minimum standards to enhance community well-being, and managing operations to eliminate the need for public subsidy and assistance programs.

As I pondered these shifts, I noticed that sustainability thinking tended to support these shifts, more or less. I knew that compared to industrial thinking, sustainability thinking often tolerates more variability, accepts lower efficiency in return for greater long-term stability and is more likely to take advantage of place-based knowledge. Sustainability thinking has been informed by and has inspired a multitude of social innovations designed to address the harms of industrialism, for example permaculture, intentional communities, life-cycle design, biomimicry, recycling, B-corps and social impact investing—just to name a few.

It is in those last two shifts on my list—regional resources and regenerative economy—that the usefulness of sustainability thinking is less clear to me. There is no question that sustainability thinking has promoted some new kinds of economic relationships in society; there are many examples in food and farming. The problem is that sustainability thinking has focused on solutions that tend to engage and promote existing economic relationships.

Mainstream sustainability thinking ultimately fails to deliver on all three rules of resilience. Yes, sustainability thinking promotes more equitable market relationships at local to global scales, but it prefers to address market inequities with charity rather than changes that shift economic power. Yes, sustainability thinking promotes energy and material efficiency and has stimulated the development of local food, local clothes, local energy and local currencies, but it still relies on a global industrial metabolism. Yes, sustainability thinking promotes the accumulation of community wealth, particularly in the form of improved environmental quality in privileged communities, but it often refuses to acknowledge the longstanding inequities that make a sustainable lifestyle possible for some.

Despite these weaknesses in sustainability thinking, all this pondering convinced me that sustainable agriculture—already on the ground and growing across our nation—might offer a head start on making the shift towards a resilient future. So I asked myself the question, "Is sustainable agriculture a resilient agriculture?"

Is Sustainable Agriculture a Resilient Agriculture?

O VER THE LAST CENTURY, the U.S. food system evolved to rely on large-scale, vertically integrated, industrial monocultures to produce, process and deliver food and other agricultural products to consumers in the United States and beyond. This dramatic transformation of the way that we eat was driven by a laser focus on land cultivation and labor efficiency with little regard for the associated damages done to land, people and community.[1] This evolution is widely viewed as a spectacular success, a story told as a modern day hero's journey. This story tells a tale of how American ingenuity, used technology and fossil fuels to win the war on hunger, release millions from the drudgery of farming and rural life, and made possible a new and better life for all in an urban utopia untroubled by the limitations of nature. It is a story built on a fragile foundation that defines cheap food as success and the production of low-cost raw materials for multinational corporate supply chains as the sole purpose of agriculture.

Despite the obvious appeal of this story to some, rural people all over the world have consistently resisted the industrialization of their food-ways, calling into question this narrow definition of success and calling out the unexamined assumptions that rationalize the exploitation of land, people and community. The multitude of environmental, social and economic harms that accompanied the industrialization of U.S. food and farming touched off wave after wave of resistance—by Indigenous, indentured, and enslaved peoples from the earliest days of colonization, through the agrarian uprisings of the nineteenth and twentieth centuries, to the most recent calls for a just transformation of our food system today.

During the oil shocks of the 1970s, this resistance to industrial food began to spread beyond rural communities and communities of color into white urban and suburban communities to emerge as the contemporary sustainable agriculture and food movement. In its early years the new movement welcomed broad discussions about the identity, structure and purpose of U.S. food and farming, but the movement was soon dominated by a focus on innovating market-based solutions to address declining farm income and rising food insecurity. These solutions were designed to connect producers with predominately white middle-class consumers who could pay premium prices for "values-added" food and to subsidize food sales to enhance food security.

By the turn of the century, many of the food justice concerns that had originally sparked the movement in rural communities and communities of color had been pushed to the fringes, particularly the longstanding social justice challenges faced by new farmers and farmers of color, farm and food system workers, and disadvantaged communities. Local food pushed aside organics to take center stage as sustainable farming organizations worked to develop new markets for local food with farmers market programs, local food guides, farm to school and farm to college programs, investments in local food processing and distribution infrastructure, and local food certification programs. Through all of this change, the fundamental goal of the sustainable agriculture movement, however imperfectly realized, remained constant: to cultivate the environmental, social and economic well-being of farmers and society.

What Is Sustainable Agriculture?

Congress created a useful legal definition of sustainable agriculture in the Food, Agriculture, Conservation and Trade Act of 1990. This definition explicitly acknowledges the multiple dimensions of sustainability—ecological, social and economic—and provides a general description of the purpose of sustainable agriculture and some desirable qualities and behaviors:

> The term sustainable agriculture means an integrated system of plant and animal production practices having a site-specific application that will, over the long term:
> - satisfy human food and fiber needs
> - enhance the environmental quality and the natural resource base upon which the agricultural economy depends
> - make the most efficient use of nonrenewable resources and on-farm resources and integrate, where appropriate, natural biological cycles and controls
> - sustain the economic viability of farm operations
> - enhance the quality of life for farmers and society as a whole

This definition of sustainable agriculture became my touchstone as I worked to understand resilience and apply it to the question of climate change adaptation in food and farming systems. While it seems pretty simple at first glance, this definition highlights some important distinctions between industrial and sustainable agriculture.

The focus on ecological health and community well-being as the basis of agricultural productivity clarifies a fundamental difference between sustainable and industrial agriculture.[2] The industrial philosophy of agriculture[3] views the production of agricultural products as no different from the industrial production of other goods. The farm is a factory, the plants and animals are workers on an assembly line. The farmer is the factory manager, tasked with managing purchased energy and materials to produce high volumes of uniform products as quickly as possible, at the lowest cost possible and with as little labor and land as possible.[4]

The resilience of the U.S. industrial food system is achieved through continuous financial subsidy in the form of publicly-funded research and development, direct payments, cost-share programs and subsidized crop insurance and disaster payments, along with additional resource subsidies acquired through the exploitation of fossil fuels and the degradation of soil, water, air quality, biodiversity and community health and well-being.[5] Paradoxically, public support for industrial food has at the same time eroded its resilience, because this support shelters the system from social and environmental disturbances.[6]

We know that food systems are more vulnerable to global disturbances and shocks when they have the following characteristics: a heavy reliance on external or distant resources; low diversity; inequitable access to resources; inflexible governance; highly specialized production, supply and marketing chains; and subsidies (both financial and technological) which mask environmental degradation.[7] These characteristics, widely recognized as the root cause of the environmental, social and economic harms of the U.S. industrial food system, take on new importance as we consider how to best to sustain American agriculture in a changing climate.[8]

The sustainable philosophy of agriculture takes a broader view of the potential contributions of agriculture to society. This view recognizes that agriculture has the capacity to produce many ecological and social goods in addition to supplying sufficient food and fiber. As the definition declares, agricultural operations have the potential to enhance (improve and sustain) natural resources, farm profitability and quality of life both on and off the farm. Agricultural operations that are most likely to realize this potential will do so through specific behaviors, for example, the use of practices that:

- manage for broad resource efficiency (not just land and labor)
- integrate crops and livestock (rather than separate them)
- cultivate ecosystem services (rather than exploit and degrade them)
- tailor practices to local social-ecological conditions (rather than overcome them)

Perhaps you are beginning to see why I started to wonder about the resilience of sustainable agriculture. It seemed to me that the desirable

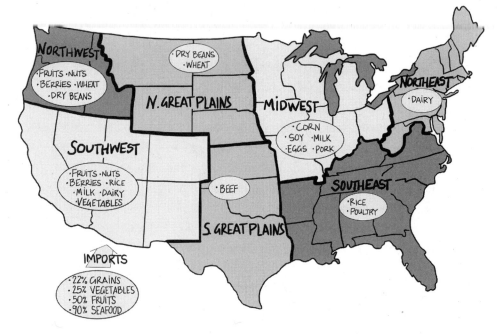

GEOGRAPHY of the U.S. FOOD SUPPLY

NORTHWEST
- FRUITS · NUTS
- BERRIES · WHEAT
- DRY BEANS

DRY BEANS
- WHEAT

N. GREAT PLAINS

MIDWEST
- CORN
- SOY · MILK
- EGGS · PORK

NORTHEAST
- DAIRY

SOUTHWEST
- FRUITS · NUTS
- BERRIES · RICE
- MILK · DAIRY
- VEGETABLES

BEEF

SOUTHEAST
- RICE
- POULTRY

S. GREAT PLAINS

IMPORTS
- 22% GRAINS
- 25% VEGETABLES
- 50% FRUITS
- 90% SEAFOOD

Figure 10.1. American plates are filled to overflowing by an industrial food system with a global reach. Regions supplying about 50 percent or more of selected foods to the domestic food supply are noted. California leads the nation in food production and processing, but specific areas in each region contribute significantly to the domestic U.S. food supply. Imports, primarily from Mexico, Canada, Chile, Central America and Asia, fill seasonal gaps in domestic production, or supply a kind or quality of food not produced domestically. Continuing drought and the unsustainable consumption of surface and groundwater supplies in the Southwest and Southern Great Plains threaten key production areas of fresh vegetables, fruits, nuts and beef. Regional production estimates were calculated for different foods from USDA NASS 2017 Annual Production Summaries. Credit: Caryn Hanna.

outcomes included in the legal definition of sustainable agriculture made a pretty good start at following the three rules of resilience:

- Diverse networks of equitable relationship are cultivated by the requirements that agricultural operations use practices with site-specific application (i.e., be place based), integrate plants and animals, make use of natural biological cycles and controls, and enhance environmental quality, natural resources, the financial viability of farms and the quality of life for farmers and society as a whole.
- Regional self-reliance is cultivated by the requirement that agricultural operations use site-specific practices that enhance natural resources, make efficient use of nonrenewable resources and on-farm resources in order to satisfy human food and fiber needs, sustain farm viability and enhance the quality of life for farmers and society as a whole.
- Community-based wealth is cultivated by the requirements that agricultural operations satisfy human needs, enhance natural resources, sustain farm profitability and enhance the quality of life for farmers and society as a whole.

Although it seemed like a pretty good start, I still had a lot of questions. Have experienced sustainable producers noticed changes in weather patterns? Have they made any changes to their operations because of changing weather patterns? Do sustainable farming practices like crop diversity and the integration of crops and livestock increase or decrease vulnerability? How does the substitution of on-farm resources for purchased inputs change farm sensitivity to variable weather and extremes? Can diversified, place-based practices sustain the economic viability of farms despite being ineligible for the public subsidies provided to industrial farms? Do sustainable producers have access to the knowledge and tools they need to manage climate risk? How do sustainable producers think about the resilience of their operation? What are the barriers to and opportunities for the development of a sustainable U.S. agriculture capable of sustaining community well-being as the pace and intensity of climate change increase in the coming years?

Will Harris has made the shift from industrial to sustainable to resilience thinking over the 50 years that he has owned and operated White Oak Pastures, a regenerative, multi-species livestock operation in southwest Georgia near Bluffton. Will continued the industrial practices he inherited from his father when he took over the farm in the 1970s, but transitioned to organic pasture-based production in 2000, and then to regenerative agriculture in 2015. Will shares these thoughts about the differences between industrial and regenerative agriculture:

"You know, most of the people that operate in this high animal welfare/high environmental sustainability farming model are younger people who came to it from somewhere else. Not many of them are like me, or only a small percentage—former monocultural, commoditized, centralized, industrialized farmers—but I'm one of them. And I'm like a reformed prostitute, you know, I've got the zeal of the convert.

"I love the debate that I get into occasionally, probably a little bit more than most, because I am one of the good old boys. They say, 'You know, the way you farm won't feed the world versus the way we farm. We're feeding the world,' and I love it when they say that, because they say, 'You just can't produce enough.'

"When they say that, I try not to smile, and I say, 'Okay, let's have that debate, but before we have that debate, I want us both to stipulate that neither farming system will feed an endlessly increasing population.' The Earth has got a carrying capacity, and once you get beyond that carrying capacity, neither one of them is going to feed the world.

"And most of them will stipulate that. They don't want to, but they eventually will. And I say, 'Okay, well, I'll go ahead then and capitulate right up front that if we're going to run out of acres first, you win. You can feed way more people than I can if acres are the only limiting factor. If we've got unlimited water, unlimited petrol fuel, unlimited antibiotics that don't create pathogen-resistance, unlimited fertilizer resources... you win.

"'But now if the limiting factor becomes water, I'm probably going to win, because I don't use as much water as you do. If the limiting factor becomes petro fuel, I win, because I don't use as much of it as you do. And if the limiting factors become phosphates and potash and these other depleting resources, I win, because I don't use as much as you.' And antibiotics, and pesticides, and so on...I win just about any way we do it other than acreage."

Cultivating Resilience with Sustainable Agriculture

A multitude of food and farming philosophies rooted in sustainability thinking are on the ground and growing throughout the United States. Innovated over the last century by food producers working to enhance social and ecological well-being both on the farm and in community, these philosophies weave traditional wisdom and scientific knowledge into contemporary expressions of sustainable agriculture.

No matter the name or the brand, these philosophies share a common set of farming practices that tend to encourage resilient behaviors in the agroecosystem. Many of these practices have been used by growers since ancient times throughout the world, for example tillage, irrigation, crop rotation, the application of ash, manure and compost, soil testing, crop and livestock selection, perennial polyculture and rotational grazing. These practices have been remembered, discovered, developed and perfected in our times by sustainable farmers and ranchers and the people who support them.

Although the many sustainable food and farming philosophies share a common set of practices, there are some distinct differences in design and management principles that are important to resilience considerations. This is because the underlying principles of a food and farming philosophy—often unacknowledged and unexamined—have the greatest power to shape the identity (structure, function, purpose) and also the operating context of an agroecosystem.

Contemporary sustainable food and farming operations tend to fall into one of five distinct philosophies: biointensive, permaculture, biodynamic, organic and climate-smart. Each of these philosophies are limited, more or less, by industrial thinking, but the practices that they share provide a good first approximation of a climate resilience toolbox for American agriculture. Because this toolbox draws from a diversity of food and farming philosophies, it offers a useful starting point for anyone in the global industrial food system who is ready to move beyond sustainability and get started cultivating resilience.

Biointensive Agriculture: Vegan, Small-Scale, Human-Powered Food Production

Biointensive agriculture is a small-scale, labor-intensive system of food production that places an emphasis on the production of extremely high crop yields through the use of deep soil preparation, composting and intensive planting. When managed as a whole system of eight complementary practices known as "Grow Biointensive," biointensive agriculture has been proven to reduce water, fertilizer and fossil fuel inputs by 50 to 90 percent and increase calorie production per unit area up to four times compared to industrial production systems. The intensive focus on deep soil preparation and on-farm compost production, combined with resource use reductions, offer significant advantages to both climate change mitigation and adaptation.[9] Ecology Action, a nonprofit organization based in California, conducts research, provides educational programs and supports the development of Grow Biointensive agriculture throughout the world.[10]

In recent years, small-scale, diversified organic growers have adapted Grow Biointensive principles to innovate a new form of biointensive agriculture that emphasizes market production and uses plastic mulches to accelerate soil decomposition processes and manage weeds, a practice known as occultation.[11] The Community Alliance with Family Farmers, a California-based organization, has an active research and education program to promote biointensive no-till organic vegetable production in partnership with Singing Frogs Farm, Bowles Laboratory at UC Berkeley and the Multinational Exchange for Sustainable Development.[12] You can learn more about biointensive agriculture practices from Hannah Breckbill and Ira Wallace and Mary Berry in this volume.

Permaculture: Earth Care, People Care, Fair Share

Permaculture is an ecological design system for sustainable human habitats as well as a global movement of practitioners, educators, researchers and organizers bound by three core ethics: care for the Earth, care for the people and care for the future.[13] Originally a contraction of the term "permanent agriculture," is now more commonly considered to mean "permanent culture" as the movement has expanded its scope since the 1970s to encompass the ecology of human settlement, economy and culture.[14] Permaculture principles include an emphasis on the place-based design of small-scale polycultures of annual and perennial plants to cultivate food, fiber and fuel. The use of perennial plants, particularly shrubs and trees, and the resource use reductions possible in polyculture systems offer advantages to both climate change mitigation and adaptation in food and farming systems.[15] Although more typically used to meet local community needs, interest in the potential of permaculture for large-scale, commercial agricultural operations is growing.

Research at the Land Institute in Kansas has focused for more than 30 years on the development of perennial grain production systems modeled on the native prairie biome.[16] The Savannah Institute supports an active research and education program exploring the potential of large-scale permaculture production systems on farms in the Midwest.[17] Carbon Harvest is a cooperative platform that facilitates the production and sale of locally grown carbon credits produced by perennial polycultures in urban, suburban and rural landscapes.[18] You can learn more about permaculture designs for large-scale agricultural operations from Mark Shepard in this volume.

Biodynamic Agriculture: The Farm as an Organism

Biodynamic agriculture is a holistic, ecological and ethical approach to farming, gardening, food and nutrition.[19] The farm is managed as an integrated, whole, living organism made up of many interdependent elements: fields, forests, plants, animals, soils, compost, people and the spirit of the place. Biodynamic farmers work to nurture and harmonize these elements by placing an emphasis on managing on-farm resources for soil, crop and livestock health and vitality through the use of crop diversity, composts, microbial inoculants and herbal remedies. The focus of biodynamic agriculture on the production of high-quality composts and the on farm production of the resources needed for crop and livestock production offers advantages for both climate change mitigation and adaptation.[20]

Biodynamic producers benefit from an international certification program, managed by the Demeter Association, which provides guidelines for soil, crop and livestock management and identifies products generated using biodynamic practices in commercial markets.[21] Fred Kirschenmann's 2,600-acre grain and livestock farm in North Dakota offers a model of a mid-sized biodynamic farm.[22] You can learn more about biodynamic agriculture from C. Bernard Obie in this volume.

Organic Agriculture: Healthy Soils, Healthy Crops, Healthy People, Healthy Planet

Organic agriculture is a production system that relies on ecosystem processes, biodiversity and local ecological cycles to sustain the health of soils, ecosystems and people.[23] Widely recognized by agriculturalists and consumers worldwide and practiced at scales ranging from small home gardens to large industrial operations, commercial organic producers benefit from national and international certification programs such as USDA's National Organic Program,[24] which provide guidelines for soil, crop and livestock management and identifies products produced with organic practices as value-added in commercial markets.

Although the climate change mitigation and adaptation benefits of organic farming have been the subject of contentious debate in recent years,[25] organic practices that enhance soil health and biodiversity have widely recognized climate resilience benefits.[26] These benefits sparked the re-emergence of regenerative organic agriculture, a term coined by Robert Rodale in 1942 to describe a holistic approach to farming whose philosophy of production strives for the continuous improvement of environmental, social and economic sustainability.[27] Regenerative organic producers benefit from a new national certification program that provides guidelines for soil, crop and livestock management and identifies products generated using regenerative organic practices in commercial markets.[28] Certified regenerative organic farming is unique within the broader regenerative agriculture movement because of its emphasis on fair labor standards for farmworkers.

You can learn more about organic agriculture from Bryce Lundberg, Albert Straus, Mark Shepard, Jordan Settlage, Pam Dawling, and Ira Wallace and Mary Berry in this volume.

Climate-Smart Agriculture: Sustainability Intensified

Climate-smart agriculture seeks to sustainably increase productivity, enhance resilience (adaptation) and avoid/reduce/remove greenhouse gases (mitigation) in order to achieve national food security and sustainable development goals.[29] This strategy, sometimes referred to by proponents as "sustainable intensification," puts an emphasis on improving the production efficiency of agriculture through a reduction in inputs (land, labor, energy, nutrients, pesticides and water) per unit crop yield. Some commonly used strategies to achieve sustainable intensification include precision technologies, diversification, substituting ecosystem services for production inputs, promoting adaptive capacity to environmental change and managing agricultural landscapes for multiple ecosystem services.[30] You can learn more about climate smart agriculture from Bryce Lundberg, A. G. Kawamura, and Albert Straus in this volume.

The intense focus on food and farming systems as both a driver of climate change and a source of solutions, plus corporate interest in monetizing carbon has sparked a number of variations on the climate-smart philosophy in recent years. Carbon-neutral agriculture seeks to reduce the carbon footprint of an agroecosystem to zero through a mix of emission avoidance, reduction and sequestration strategies.[31] Carbon farming seeks to remove CO_2 from the atmosphere and store it in plant material and/or soil organic matter at a rate that exceeds carbon losses.[32]

These kinds of production philosophies can be helpful; however, their potential value to agricultural resilience is limited by the extreme focus on carbon reduction without considering the broader operating context. This kind of thinking—a hallmark of industrialism—serves to protect the status quo and often leads to textbook maladaptation. A good example is the current push for public subsidy of regional biogas development in North Carolina's pork industry despite its vulnerability to climate change. The industry operates in a region with a high risk of damaging hurricanes and river flooding, uses production practices that are known to increase food system vulnerability to global disturbances and shocks,[33] and has degraded the adaptive capacity of the region through generations of well-documented exploitation of coastal plain communities.[34]

A third kind of climate-smart agriculture has captured the imagination of the public in recent years. Regenerative agriculture has no legal or regulatory definition nor has a widely accepted definition emerged in common usage.[35] In general, regenerative agriculture philosophies strive to generate multiple benefits associated with improved soil health in agroecosystems through the adoption of practices that increase soil carbon sequestration and produce marketable carbon credits. The philosophy differs markedly with regard to desired benefits. For example, General Mills has pledged to invest in regenerative agriculture practices on one million acres of farmland by 2030 to enhance the resilience of their supply network.[36] In contrast, the Regenerative Agriculture Alliance is working to establish a "regional regenerative poultry corridor" producing meat and eggs on 150,000 acres in the central Midwest that "delivers equitable returns, just wages, dignity for workers, and ecological regeneration."[37] You can learn more about regenerative agriculture from Gail Fuller, Rebecca Graff and Tom Ruggieri, Walker Miller, and Jamie Ager in this volume.

Although these five distinct philosophies of sustainable agriculture can serve as a foundation for our resilient food future, they will not get us all the way there. Sustainability thinking, although an important step forward, is still rooted in the myths of industrialism. Sustainability thinking promotes the belief that economic efficiency is the basis of community well-being, that we can produce a limitless supply of natural resources, that we can maintain a continuous global flow of energy, money and materials, and that our climate is stable. Sustainability thinking fails to fully acknowledge the harms of industrialism. Although a measurable improvement over industrial business as usual, sustainability thinking is simply not up to the challenge of guiding the transformation to a resilient agriculture.

Resilience thinking goes beyond sustainability by offering new concepts and tools that can help us let go of the limits of "business as usual" industrial thinking. It gives us a common language and a new framework for making decisions that we can use to envision and shape change together to cultivate a resilient future for land, people and community into the twenty-first century. This is the essential difference between sustainability and resilience.

11

Resilient Agriculture: New Tools for Shaping Change

S O HOW DO WE PUT resilience thinking to work on the ground and in the real world? What can we take from sustainability thinking to help us make the shift from industrialism to resilience? As I explored the answers to these questions, I learned that natural resources managers have been hard at work for decades applying resilience thinking to restore and protect the health and well-being of natural and working lands in the midst of global change.

These natural resources managers have already made the shift from sustainability to resilience thinking through the use of management practices that cultivate regular opportunities for the people in the agroecosystem to practice reflective learning.[1] Recall that reflective learning promotes place-based innovation with learn-as-you-go practices that can be used by all producers—no matter their farming philosophy—to enhance the response, recovery and transformation capacity of agroecosystems.

Navigating Uncertainty with Adaptive Management

All farmers and ranchers manage a complex network of constantly changing social-ecological relationships to achieve a desired purpose. Sustainable producers strive to generate multiple social and ecological benefits—sustainability benefits—through the use of sustainable practices that are well-aligned with their management goals and operating context. Resilient producers take on an additional challenge: they strive to follow the three rules of resilience (cultivate reciprocal networks,

150

regional self-reliance and the commonweal) in order to promote the resilience of their operation. Adaptive management can help.

 Mark Frasier draws on many resources to enhance his beef operation's capacity to weather variability and extremes. "I look at that key word, variability," Mark says. "You've got to have adaptive management to respond, both in terms of knowing how to respond, but also anticipating what a change will bring. Oftentimes making a timely decision is key, either in a cost-saving sense or in a sense of conservation of natural resources."

⌒◯◯◯◯ ◯◯◯

Adaptive management enhances the resilience of a system by providing the language and some simple tools that help managers formally evaluate the performance of that system relative to its purpose. We all use a kind of adaptive management informally in our daily lives when we set a goal (e.g. lose weight, save for a vacation or earn a new job certification), make plans to support this goal and act on those plans, check our progress towards the goal (e.g., number of pounds lost, the increase in savings, or progress towards certification) and make adjustments as needed to achieve our goal. These are the four steps of adaptive management (set goal, plan and act, monitor progress, adjust as needed to make progress towards goal). Notice that these steps support the kind of engaged "learn-as-you-go" thinking that enhances the resilience of social-ecological systems. Once you know about these four steps of adaptive management, you will likely start to notice how often we use them, even if only informally, and how these behaviors contribute to our well-being. Used with a bit more attention and regularity, these behaviors have a unique transformative power that has proven successful in social-ecological systems of all kinds including individuals, families, businesses and communities.[2]

A key feature of adaptive management is regular monitoring of system behavior over time, so that management strategies can be adjusted to enhance the system's ability to achieve its goals as conditions change.

By tracking the successes and failures of different adaptive actions, adaptive management can help producers evaluate how different farming practices support progress towards their goals. Adaptive management offers farmers and ranchers a valuable set of tools to identify "best fit" practices to enhance the resilience of their operations and to tailor these practices to the on-the-ground realities of their operations.

Developed more than 50 years ago to promote the sustainable management of natural resources, contemporary adaptive management has more recently been applied to sustainable business and community development, including sustainable farm management. Adaptive management principles emphasize active learning about the system under management through a continuous process of evaluating and adjusting management actions based on regular observation of system behavior.[3] Because adaptive management is particularly useful in conditions of high uncertainty and complexity, it has been recognized as a valuable tool for enhancing the climate resilience of U.S. farms and ranches.[4]

Whole Farm Planning Is Adaptive Management

Contemporary whole farm planning was originally developed to reduce the environmental impacts of industrial agriculture, but it very quickly expanded to include sustainability considerations.[5] Whole farm planning makes use of the "adaptive management cycle," an iterative process involving five steps: goal setting, resource assessment, planning and implementation, monitoring progress towards goals and replanning.

Nearly all agricultural producers use some kind of performance indicators to monitor the performance of their operations and guide business decisions. Yield and income are two very common indicators of performance used by all agricultural producers.[6] Sustainable producers add some additional performance indicators to achieve their social and environmental goals, for example, the change over time in their need for fertilizers, pesticides and irrigation.[7] Keeping all of these goals in mind while making short-term production and marketing decisions can be challenging. Whole farm planning provides a strategy and some easy-to-use tools to aid agricultural producers in managing for the multiple goals of sustainability and resilience.

Pam Dawling appreciates the information gained through regular monitoring of soil and crop conditions on her vegetable farm in Virginia, especially as weather patterns have become more erratic. "I've been doing more things like recording soil temperatures because I think it will make a big difference in knowing when to plant things, rather than relying on calendars to guide field work." Other variables that Pam monitors daily or weekly on the farm include temperature, precipitation, the seasonal timing of plant and animal development such as leafing, flowering or fruiting (phenological benchmarks), and pest and disease populations. "When I do my once a week walkaround, I am sure to go everywhere and look at everything to be sure there are no surprises about to burst on the scene. I think this is very worthwhile, because it is so easy to get focused on urgent tasks and lose sight of the important ones. To be a successful farmer, you've got to find a way to balance the importance and the urgency."

∽◌◌∽ ◌◌◌

Monitoring production system performance with sustainability indicators[8] can improve farm and ranch business management in several ways. Indicators can help to clarify goals and assess options, and can be used to evaluate the success of changes in farm practices. Monitoring can be particularly useful as an early warning sign that a change in management has been successful or is not going as planned. The monitoring step in whole farm planning is likely to become more important in agricultural risk management, no matter the production philosophy, as changing weather patterns, coupled with other disturbances and shocks associated with climate change, increase the uncertainty and complexity of agricultural production.

Farmers who have adopted whole farm planning feel more confident in their management decisions and report improved profitability, enhanced quality of life and increased natural resource quality on their farms and ranches.[9] Although whole farm management is a proven sustainable management tool, it is not widely used at present, even by sustainable farmers. Many farmers say they are reluctant to use these practices because of the recordkeeping required for effective monitoring;[10] however, whole farm management is increasingly recommended as a sustainable management practice and is a key feature of many beginning farmer training programs.[11]

 Business planning at Humble Hands Harvest in Iowa is grounded in the holistic management practices that co-farmers Hannah Breckbill and Emily Fagan both learned in farmer training programs. "Farm Beginnings starts with Holistic Management," Hannah explains, "and I really appreciate that way of looking at things. It has been very valuable to me. We have a mission and vision statement that we go back to and we have communication processes that we feel are really important."

∾◌◯◌◌ ◯◌◌◯

Many of the farmers and ranchers featured in *Resilient Agriculture* use holistic management to enhance the response and recovery capacity of their operations and to transform their operations when circumstances require it. Ranchers Julia Davis Stafford and Mark Frasier manage large-scale cow-calf operations in New Mexico and Colorado with holistic practices. Rancher Gary Price used holistic management to restore native grasslands on his ranch in East Texas with planned grazing. Faced with catastrophic weather damage, North Dakota rancher Gabe Brown used holistic management to transform his industrial grain operation into a highly diversified and integrated operation producing a multitude of food grains and livestock products for local and regional markets. Holistic management helped Kansas grain farmer Gail Fuller guide his family farm through not just one, but two major transitions that included diversification, downsizing and finally moving to a new location to get out of a floodplain. Iowa vegetable growers Hannah Breckbill and Emily Fagan have a holistic goal that keeps their cooperative farm on the path to success, as they have defined it.

Two U.S.-based organizations offer tips, guides, producer success stories, consulting and training in holistic management for land managers. The Savory Institute promotes the regenerative management of improved pastures and native grasslands by conducting research, supporting land managers and influencing markets.[12] Holistic Management International provides tools and training for food and farming businesses to envision and realize healthy resilient lands and thriving communities.[13]

Nature-Based Solutions: Cultivating Healthy Ecosystems for Land, People and Community

Nature-based solutions are climate change adaptation strategies designed to enhance the climate resilience of land, people and community through the use of sustainable management practices designed to maintain, restore and protect ecosystem health. Nature-based solutions use a wide range of strategies including ecosystem-based adaptation and disaster risk reduction, green and blue infrastructure (plants and water in urban landscapes), landscape restoration and natural climate solutions to take advantage of the sustainability benefits produced by healthy, biodiverse and resilient social-ecological systems. For example, nature-based solutions are proven to support climate change adaptation through flood protection, urban cooling and the regulation of soil, water and air quality, while contributing to climate change mitigation, food security and climate justice.[14]

Many sustainable agriculture practices are nature-based solutions, even though most of these have been developed without climate mitigation or adaptation in mind. Livestock producers who use managed grazing to improve forage production in their operations also reap the climate protection benefits of healthy soils that capture more water during heavy rains and store that water for use by crops in dry periods or filter and release it to regenerate groundwater and surface water supplies. Vegetable, fruit and nut growers who mix annuals and perennials across a landscape to enhance energy flow, mineral and water cycles, and community dynamics (interactions among and between species on the farm) also reap the climate protection benefits of shade and less wind. Grain growers who integrate livestock with annual and perennial crops to reduce fertilizer and pesticide needs also reap the climate protection benefits of healthy soils.

As an additional plus, many nature-based solutions also deliver greenhouse gas mitigation benefits. For example, crop rotations that mix annual and perennial crops not only reduce the need for synthetic fertilizers and pesticides (and so reduce emissions associated with these inputs), they also increase the amount of carbon drawn out of the atmosphere by soils and plants on the farm. Ecosystem-based sustainable agriculture practices also provide other benefits that extend beyond the farm gate to improve community well-being. For example, improved soil health on the farm enhances regional water quality, decreases the need for irrigation and reduces the risk of flooding in downstream communities.[15]

Mark Shepard designed a diverse agroforestry system based on the ecological biome native to his location in southern Wisconsin. "We got started by selecting perennial plants that mimicked the oak savanna plant community that we could sell, feed to an animal or eat ourselves," Mark recalls. "Nature has been here with these species for a long time—anywhere between six thousand and nine hundred million years," Mark explains. "That's what really matters. From my perspective, the best model I have for a system that can survive all the changes, whether it's short-term weather or long-term climate, is the plant community type that I select. Then all I've got to do is figure out how to dance with that going forward."

Bryce Lundberg is upbeat about the future of rice production at Lundberg Family Farms in California. "We're a resilient family and a resilient region," Bryce says. "We're confident that we'll keep farming rice. I see no reason we can't." He points to Lundberg Family Farms' newest acquisition, Dos Rios Farm, as a good example of what is possible when a business strives to "leave the land better than you found it." Located at the confluence of the Sacramento and Feather Rivers, the new farm is a prime site for agroecosystem restoration and management designed to benefit salmon and migratory birds, while simultaneously addressing flood management concerns and supporting local organic agriculture. "Farming inside the levee is a whole new experience for us," says Bryce. "The farm is located where the North State rivers come together and it's the first farm that floods. It's very unique…we think it's the best farm in the state to grow salmon on in the winter time."

∽◌◌∽ ◌◌◌

Nature-based solutions have proven to be a flexible, cost-effective and broadly applicable strategy for climate change adaptation across a variety of landscapes and land uses throughout the world.[16] For example, coastal ecosystems such as salt marshes and barrier beaches provide natural shoreline protection from storms and flooding; urban green spaces reduce the urban heat island effect and improve air quality; restored wetlands sequester carbon, contribute to a healthy regional hydrologic cycle and reduce inland flooding; and afforestation with native species facilitates the adaptation of forests to climate change.

Long recognized internationally as a sustainable and resilient twenty-first century development strategy,[17] nature-based solutions have received little attention in the United States until recently. Some notable examples of new nature-based solutions in U.S. agriculture include corporate supply chain programs such as the Ecosystem Services Market Consortium and Indigo Ag's Terraton Challenge, public-private partnerships such as Natural and Working Lands Initiative of the U.S. Climate Alliance, non-profit programs such as the North American Climate-Smart Agriculture Alliance, the Cornell Climate-Smart Farming Program,

The Savanna Institute, the Soil Health Academy and new federal, state and local projects and programs seeking to promote nature-based agricultural climate solutions.

The Adaptive Continuum: Protect, Adapt, Transform

The adaptive continuum is a resilience thinking tool developed by natural resource managers as a guide for selecting a mix of adaptation strategies designed to both reduce the risks of specific threats and also enhance the general resilience of the resource under management.[18] This new approach to adaptive resource management has been recommended for use in nature-based agricultural adaptation,[19] has been used in U.S. federal adaptation planning,[20] and is used to categorize adaptation options in the USDA Climate Hub's Adaptation Workbook.[21]

The adaptive continuum describes a range of adaptation strategies that vary from adjustments that reduce climate risk with little or no change in agroecosystem relationships to adjustments that shift the system to a new identity. The adaptive continuum helps to clarify the full range of adaptation strategies available to agricultural managers by organizing them according to what the manager hopes to achieve with the strategy (management intention) and the effect of the strategy on agroecosystem relationships. Resilient systems typically use a mix of all three strategies that changes over time as conditions change.

Thinking about managing climate risk in this way—as a choice between a range of strategies that involve increasingly greater changes in the structure, function and purpose of the agroecosystem (its identity)—can help producers, and those that support them, identify adaptation strategies that are a good fit with management goals and resources. This way of thinking about adaptation strategies is well-aligned with both whole farm planning and nature-based solutions.

Resilient farmers and ranchers use all three adaptive strategies in a comprehensive, nature-based approach to climate risk management.[22] Keeping the whole of their operation in mind, resilient producers choose a mix of strategies with three goals in mind: to protect their operations

ADAPTIVE STRATEGY

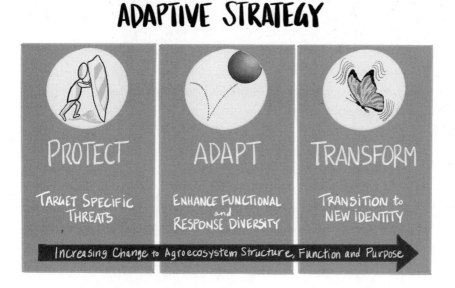

Figure 11.1. The adaptive continuum is a resilience thinking tool developed by natural resource managers to guide the selection of adaptation strategies. The continuum describes a range of adaptation strategies that vary from adjustments that reduce climate risk with little or no change in agroecosystem relationships to adjustments that shift the system to a new identity. Resilient producers strive to manage a dynamic mix of all three strategies as part of an adaptive, whole farm management plan. Credit: Caryn Hanna.

from specific, high-risk threats; to adapt more generally to threats that are less damaging, less likely or unexpected; and to prepare for likely challenges ahead.

PROTECT

Sustainable practices that serve to protect the agroecosystem from damaging disturbances and shocks without change to existing relationships—referred to here as PROTECT strategies—typically involve investment in physical and financial resources that target specific risks and provide financial resources for recovery. Some key strategies in use by farmers and ranchers in the U.S. to protect their operations involve changes in equipment and infrastructure. For example, many farmers are increasingly challenged by more variable and shorter windows in which to complete the seasonal fieldwork required for crop production. Many of these farmers have managed this new risk by purchasing more or larger equipment (tractors and associated implements) to reduce the time needed to complete fieldwork and increase the management flexibility of their operations.[23] Many producers are also adding or upgrading the infrastructure needed to physically protect crops and livestock from weather extremes, for example, adding or upgrading irrigation, drainage and protected growing space to reduce the risks associated with more frequent and intense rainfalls and drought.

Some producers are taking advantage of federal and state agricultural support programs to reduce climate risks to their operations.[24] Producers growing eligible crops and livestock, primarily commodity crops like corn, soybeans, wheat, cotton and rice, can take advantage of subsidized insurance against production risks, cost-share programs for infrastructure improvements, low-cost loans and disaster relief payments. Unfortunately, these kinds of PROTECT strategies tend to cultivate the fragility of industrial agroecosystems.[25]

Like industrial producers throughout the United States, the sustainable farmers and ranchers featured in *Resilient Agriculture* employ strategies to protect their existing operations. Nash Huber purchased additional equipment to speed the completion of his fieldwork. Jacquie Monroe expanded her irrigated production area and upgraded to more efficient irrigation systems. Jim Crawford has increased pesticide use and physically protects his fields from excessive spring rainfall with plastic mulch. Hannah Breckbill added a hoop house to increase protected growing space. Steve Ela purchased additional wind machines to enhance frost protection in his orchard. Ira Wallace and Mary Berry are shifting to no-till transplanting with a human-powered no-till planter imported from South Africa. Jim Koan increased the capacity of field drainage systems. Jim Hayes built new sheltered space for livestock. Liz Henderson added new wells and irrigation and shifted field work to cooler early morning hours. The Rid-All partners added backup generators and high-performance insulating plastic coverings to their greenhouses.

Adaptive strategies that protect existing operations—referred to here as PROTECT strategies—offer some important and unique benefits to climate risk management, but they are most effective as part of an integrated strategy that includes actions taken to promote ecological and social response capacity. This is because adaptations designed to simply protect the existing agroecosystem—especially those that are already fragile—can be costly, will likely increase in cost and decrease in effectiveness over time, and may ultimately fail as the pace and intensity of climate change increases.[26] Adaptive strategies that protect the existing operation are most effective when they are used as part of an adaptive whole farm plan to buy some time while actions taken to address resource limitations or enhance the production of ecosystem services are taking effect.

ADAPT

Adaptive actions taken to reduce climate risk by changing existing agroecosystem relationships—referred to here as ADAPT strategies—are those made with the goal of improving functional and response diversity in order to enhance ecological and social self-regulation. These actions typically involve investments in natural, social and human resources made with the expectation of generating broad climate risk benefits that enhance the general resilience of the agroecosystem. For example, practices that promote soil health, such as diversified crop rotations and the integration of livestock into cropping systems, not only buffer the agroecosystem from extremes of moisture, but produce additional risk management benefits by spreading production risk across a wide range of crops with differing sensitivities to seasonal variations in weather, soil fertility, pests and diseases. These practices also support more diversified marketing. Other sustainable agriculture practices with proven climate risk management benefits include conservation tillage and cover cropping, the intercropping of annuals and perennials, management intensive grazing and intentional management of the whole farm land-scape, including field edges, woodlands, riparian areas and wetlands.[27]

ADAPT strategies employed by U.S. industrial producers include the use of drought-resistant species and crop varieties and changes in man-agement to promote soil quality such as conservation tillage, the use of cover crops and other practices that increase the quantity and quality of soil organic matter.[28] Industrial corn and soybean producers also recog-nize the climate risk benefits of more diversified cropping systems and identify diversification as a key adaptation strategy as climate change ef-fects intensify, along with purchasing more crop insurance (a PROTECT strategy) and going out of business (a TRANSFORM strategy).[29] Recom-mended practices for livestock producers working to adapt to climate change include obtaining the education and training required to under-stand and manage animal needs, potential stress levels and options for reducing stress, and selecting animals and management strategies com-patible with local production conditions.[30]

The sustainable producers featured in *Resilient Agriculture* offer many examples of the climate risk management benefits associated with healthy soils, diversified production systems, value-added processing and high-value diversified marketing. For example, Alex Hitt shifted crop production out of the intense heat and drought of mid-summer. Jim Crawford and Richard DeWilde shifted production out of bottom lands as flooding risks have increased. Hannah Breckbill adapted to "soggy" spring weather by shifting to no-till raised beds. Kole Tonnemaker, Steve Ela and Walker Miller increased crop diversity to spread production risks in more variable weather, to take advantage of longer growing seasons and to respond to consumer demands in direct markets.

Nash Huber, Gabe Brown and Russ Zenner use dynamic crop rotation to select specific species or cultivars in each phase of the rotation best suited to current and expected weather conditions throughout the grow-ing season. Gabe Brown and Gail Fuller developed incredibly diverse cover crop cocktails to assure good cover crop production despite variable soil, microclimate and weather conditions. Tom Trantham and Gail Fuller use a pasture-cropping system that features the no-till planting of annual forages into permanent pasture. Nash Huber and Bob Quinn select for crop species and cultivars that are best suited to their local conditions and more robust to more variable weather and extremes. Gabe Brown, Will Harris, Ron Rosmann and Greg Gunthorp select for livestock that perform well in their agroecosystems.

Livestock producers Julia Davis Stafford, Mark Frasier, Gary Price, Gabe Brown, Gail Fuller, Dick Cates, Ron Rosmann, Jim Hayes, Tom Trantham, Jason Settlage and Will Harris enhance ecological response capacity by managing herd size and movement through improved pastures and annual forage crops and/or native grasslands using intensive grazing and holistic management strategies. Many have reduced herd size, leased additional grazing acreage or done both in response to more variable weather and extremes. Greg Gunthorp uses standing corn as a sheltered area for pigs and encourages mulberry trees in wooded field edges for shelter and supplemental feed.

Although ADAPT strategies offer effective risk management options to agricultural operations and produce additional benefits on farm and beyond, they are also likely to increase in cost and decrease in effectiveness as climate change effects grow more intense in coming years. Over the last decade, several *Resilient Agriculture* producers have told stories of ADAPT practices that have been overwhelmed by extreme weather events. For example, back-to-back thousand-year floods in 2007 and 2008 at Harmony Valley Farm in Wisconsin caused a million dollar loss. An unprecedented late spring freeze in Michigan caused a 90% loss in the apple crop at Almar Orchards and Cidery in 2012. Hurricane Michael caused 1.4 million dollars in damages when it hit White Oak Pastures in Georgia in 2018. Early fall freezes in 2019 and then again in 2020 at Ela Family Farms in Colorado killed all the farm's cherry trees and bearing peach trees, caused total crop loss in pears and significantly reduced apple yields.

TRANSFORM

Adaptive strategies that create fundamental changes to the identity (structure, function and purpose) of the existing agroecosystem—referred to here as TRANSFORM strategies—are a special case of adaptive actions. TRANSFORM strategies are used by managers with the intent to shift the operation into an entirely new identity that is more resilient to current and expected climate risks. TRANSFORM strategies significantly alter social-ecological relationships within the agroecosystem in order to enhance response, recovery and transformation capacity. Some common examples of TRANSFORM adaptations are substantially increasing crop biodiversity, integrating livestock into cropping systems or shifting from annual cropping to pastured livestock production.

Like ADAPT strategies, TRANSFORM strategies typically produce broad risk management and resilience benefits both on farm and in the local community. For example, managing the transformation of a conventional row crop operation in a floodplain to a managed grazing operation reduces the sensitivity of the operation to flooding, enhances regional water quality (because perennial pastures typically have lower rates of soil erosion than row crops) and can also reduce the risk of downstream flooding.[31] The time to consider the transformation of an agroecosystem is when the system, in its current configuration, fails to achieve its purpose; when the costs (ecological and social) of maintaining it become unacceptable; or when it becomes clear that a transition is inevitable because of changing environmental, social, cultural or economic conditions.

Sustainable agriculture offers a diverse set of practices that can be used to transition industrial operations to new, more climate-resilient identities. Sustainable farmers and ranchers in every region of the U.S. have transformed industrial operations using a diverse group of ADAPT and TRANSFORM strategies that are proven to generate multiple sustainability benefits. Crop diversification, annual and perennial intercrops, integrating livestock into crop production systems, adopting intensive grazing and restoring native grasslands, wetlands and riparian areas are just a few examples of sustainable agriculture practices that enhance agroecosystem climate resilience without compromising yields.[32] These practices also have the added benefit of increasing the resilience of the agroecosystem to many other kinds of environmental, social and economic disturbances and shocks.

Sustainable agriculture practices that can be used as TRANSFORM strategies for livestock production systems include a transition to livestock species or breeds that have greater tolerance of changing climate conditions, forages and grassland quality, pests and diseases, and the use of multispecies production systems (for example, following cattle with poultry). The conversion from industrial livestock production in confined animal feeding operations to integrated crop/livestock production systems is a TRANSFORM strategy that has been recommended to reduce detrimental environmental and social impacts, improve the profitability and sustainability, and enhance resilience to climate change in U.S. livestock production systems.[33]

Some of the producers featured in *Resilient Agriculture* offer compelling examples of the power of transformation as a climate change adaptation strategy. Brendon Rockey transformed an industrial potato production system by restructuring crop diversity to enhance water conservation. Mark Shepard transformed a degraded Midwest corn and soybean farm into a diversified perennial polyculture. Jordan Settlage transformed a Midwest corn and soybean farm into an organic pasture-based dairy. Gabe Brown used holistic management to guide the transformation of an industrial grain production system and degraded native prairie into a sustainable and resilient diversified crop and livestock production system. Greg Gunthorp transformed a pastured hog production system into a diversified livestock production system with an on-farm processing plant and retail store. Will Harris transformed his family's cattle feeding operation into a regenerative multispecies livestock operation with an on-farm processing plant and retail store. The Rid-All Green Partnership transformed a burned-out Cleveland neighborhood into a thriving urban farm. Walker Miller transformed an old, worn-out southern hill farm into a thriving diversified U-pick fruit operation.

These stories, along with many more stories of food and farming system transformation told by others over the last decade, provide just a glimpse of the transformative power of sustainable agriculture. We already have much of the knowledge and most of the tools that we need to make the shift from industrial to resilient agriculture. A resilient food future is not only possible, it is already on the ground and growing throughout our nation.

Barriers to a Resilient Agriculture: From the Farm Gate to Your Plate

In recent years, farmers and ranchers have moved to center stage of the climate change story. Presented as heroes by some, villains by others, the truth of the matter is more complex. Yes, there are powerful climate change solutions in agriculture. In fact, according to one kind of accounting, some of the most cost-effective climate solutions are found in food and farming.[34] Yet, despite the long experience of successful sustainable farmers and ranchers growing food in every region of this country, very few U.S. producers take advantage of sustainable practices that promote climate resilience. What's stopping them? It turns out that it is us!

Recall that the adaptive capacity of an agroecosystem is determined by the operating context, willingness of the producer to act and access to knowledge and resources for adaptation. Farmers and ranchers in the U.S. are making decisions in an operating context that creates many significant barriers and very few opportunities to cultivate climate resilience. Some barriers can be found on the farm and are under the producer's control, but many are created by social and ecological conditions that are created by society—by government, the agrifood industry and consumers.[35] As an example, consider this comparison of the challenges encountered during the transition to more climate-resilient agroecosystems at Rockey Farms and Fuller Farms.

Brendon Rockey transformed his potato farm from an industrial agroecosystem (emphasizing financial and physical assets) to a more resilient agroecosystem (emphasizing natural, social and human assets) through fundamental changes in crop diversity, soil health and management perspective—all components of the agroecosystem under his management. Brendon recognized that the agroecosystem had to change (water supply was dwindling), he was aware of practical alternatives (soil-building cover crops) and he had access to all the assets he needed to successfully guide the transition. The barriers presented by the operating context were primarily on-farm, although Brendon says he did have to put up with a lot of ridicule from his neighbors and friends during the early stages of the transition.

Gail Fuller's transition from an industrial corn and soybean farm to a diversified grain and livestock farm was hampered by a lack of information about diverse crop options and conflicting federal crop insurance program regulations. Gail recognized the need for change (poor soil health) and was aware of practical alternatives developed in other regions (cover crop cocktails and increased diversity of cash crops), but he could find no information about these practices for his region—he had to mostly just figure it out for himself. In addition, conflicting federal program requirements restricted Gail's ability to innovate new crop practices without giving up the recovery capacity provided by government-subsidized crop insurance. These barriers created significant challenges to Gail's ability to manage climate risk. Eventually, with a heavy debt load and increasing flood risks threatening his ability to stay in business, Gail made the tough decision to sell his family farm.

As another example of how society creates barriers to resilient innovation, consider how new federal regulations designed to reduce food safety risks in industrial agroecosystems threatened Jim Koan's innovative fruit silvopasture system at Almar Orchards and Cidery. Jim recognized the need for change (current farm practices violated new food safety regulations), discovered a credible alternative (sell hard cider instead of fresh apples) and had access to all the assets he needed to achieve the transition. In this case, the agroecosystem had sufficient resilience to innovate a solution to changing regulatory conditions, plus the shift to shelf-stable products offered unexpected climate resilience benefits as frost damage in spring has grown more frequent and intense over the last decade.

Although federal and state government efforts to support climate change adaptation in agriculture and other sectors of the economy have accelerated over the last decade, these efforts tend to promote strategies that protect existing industrial agroecosystems rather than strategies that promote social-ecological response and recovery capacity. This choice to protect industrial agriculture despite its many well-documented harms to land, people and community is a classic example of maladaptation, because adaptation investments are being wasted to protect business as usual rather than to encourage a shift to agroecosystems with the social-ecological resilience to sustain community well-being over the long term.[36]

 Dairy farmer Jordan Settlage doesn't use practices on his Ohio farm that, in his view, promote death, which he sees as the primary operational mode of industrial agriculture. "We try to keep away from all the 'cides' that are rampant in conventional agriculture: pesticides, insecticides, fungicides, herbicides, homicide and suicide." He goes on to explain, "The suicide rate is out of control in farming. Farmers are killing themselves at a rate like they were in the '80s, and on top of that, you've got farmers shooting each other over dicamba (a new long-lasting herbicide) drift. How can we look at that and not see that maybe that's a problem?"

ⱣⱣⱣⱣ ⱣⱣⱣ

We know a lot about how to design and manage agricultural operations with the capacity to enhance the sustainability and resilience of our nation's food supply. Many U.S. farmers recognize they need to change how they produce food. They also know about the kind of changes that are needed—they are well aware of the "proof of concept" innovations made by sustainable producers across the diversity of North American landscapes. The simple truth is that the operating context of U.S. agriculture creates barriers to the ability of farmers and ranchers to self-organize, learn, innovate, adapt and transform their operations in response to changing conditions.

Managing for sustainability and resilience at the individual farm and ranch scale has been constrained by cultural forces that have favored the industrialization and globalization of the U.S. food system for more than 300 years. These forces strive to protect the existing system despite the fact that it has not fulfilled its purpose (save the family farm, feed the world and provide Americans with a safe, abundant and affordable food supply) and there is overwhelming evidence that it is not sustainable or resilient in the face of twenty-first century challenges. Cultivating a resilient agriculture cannot be accomplished through actions taken only by producers in their own operations. A resilient agriculture requires a transformation of the operating context of American agriculture—the U.S. industrial food system.

PART 3

WHAT PATH TO RESILIENCE?

The new dawn blooms as we free it.
For there is always light,
if only we're brave enough to see it.
If only we're brave enough to be it.

— Amanda Gorman, U.S. Youth Poet Laureate,
Presidential Inaugural Poem, 2021

The Light and the Dark
of These Times

As I write this in the summer of 2021, I pause to think back on all that has happened in the 18 months since I started working on this book. Can it really have been little more than a year ago that we heard the first news about the coronavirus and then watched it spread like wildfire around the world? The pandemic hit New York City hard and not long after all across our nation our "good life" began to tremble, bend and finally break under the stress of shutting down our communities and sending many of us home. Almost overnight, COVID split us into a society of just two kinds of people—those that deliver and those that are delivered to.

I can't help but see both the light and the dark of these times. So much disruption of our lives, so much loss, so many of our most deeply held beliefs called into question; and so many lessons learned. This has been a time of terrible loss and improbable triumph as the pandemic transformed how we work, how we teach and how we eat. These unprecedented times invite each of us to step back and consider what has been lost and what has been gained in our pursuit of the American Dream. These times shine a bright light on a 300-year journey marked by the relentless exploitation of land, people and community. They invite us to take an honest look at our own participation in this journey and reflect on how it has shaped the reality of our lives today.

This is a demanding invitation. It is not easy to confront the suffering of land, people and community that makes our good life possible.

173

It's not easy to recognize our own complicity in this suffering and it is even harder to imagine what we can do about it. There seems to be a lot of evidence that we—as a people, as a nation, as a culture rooted in colonialism—may be incapable of change. We have been living a story that celebrates our determined march out of nature and into the city. This story encourages us to deny the damage done along the way. It is a story that holds on tight to the belief that technology will save us—that technology can overcome all social-ecological limits to growth. This is the story that has pushed our planet to the threshold of collapse.

This story has been so successful because it feeds the dark side of human nature: the side that is greedy, that always wants more, that fears being left out, less than, left behind. The side that believes competition is the only path to success. It is human nature to discount the future, to favor our tribe over others and to live in denial of the damage done. We are biologically wired to focus on achieving short-term individual and tribal goals.

But human nature has a light side too—the side that is generous, that always has enough, the side that embraces cooperation in community as the path to success.[1] This too is our story. It is our legacy, it is our light, it is a story rooted deep in the Indigenous wisdom that recognizes that we are all one—that the health of land, people and community are inextricably connected.

The question is, how do we step into this light? We can start by having the courage to imagine the more beautiful world that we know is possible. To envision this future and then to imagine how you might work with others in your community to cultivate a new way of feeding ourselves. A way that has the power to heal the wounds of industrialism and sustain the well-being of land, people and community.

The good news is that we don't have to imagine this path, or be the first to step onto it. There are already many people walking this path—you've already met some of them in this book. They are part of a lively community of farmers and ranchers, farm and food system workers, researchers, policymakers, activists and eaters who are drawn to the idea of sustainable food. You can step on this path too. You can see it, hear it, smell it, touch it and, best of all, taste it at your local farmers market or cooperative grocery or local food restaurant, by volunteering in a community garden or feeding program, or visiting a sustainable farm or food processing business in your community.

What if we could rewrite the climate change story, from one of loss and regret, to one that tells the story of humanity's journey back into the light? What if we take action today, so that one day we can look back and tell our children about the time that we decided to change the way we eat?[2] What does a resilient national food system look like and what will it take to get us there? Just as sustainable farmers and ranchers offer us glimpses of a resilient agriculture, the lessons learned in the sustainable food movement shine a light on what it will take to make the transition from industrial to resilient foodways.

A resilient U.S. food system will rest on a national network of regional foodsheds, each embedded in a particular place defined perhaps by bioregion,[3] or watershed or metropolitan area. This New American Food System will support diverse local and regional networks of mutually beneficial and reciprocal relationships that produce fresh, nutrient-dense foods along with a multitude of other ecological and social benefits. These regional foodsheds, each a unique expression of land, people and community, will have the ecological and social diversity needed to promote regional self-reliance and accumulate the wealth needed to heal the wounds of industrialism and sustain community well-being over the long term.

Resilience emerges through the diversity of linkages within and across multiple scales, from the local to the global, and by applying ecological patterns and adaptive management strategies to the design and development of sustainable foodways. Agricultural inputs, production, processing, distribution and sales are focused on regional markets and food trade is limited to a few specific foods that are particularly well-adapted to the region, for example almonds in California and wheat in Montana.

Taking a cue from nature, each regional foodshed will produce its own renewable energy and recycle its wastes to close regional materials cycles. These foodsheds will likely be less labor efficient and possibly less land efficient, but they will reside in a sweet spot—an agroecological "window of vitality"—that cultivates the ecological and social response, recovery and transformation capacity that can sustain all land, people and community over the long term.

The stories that I've shared in the first and second editions of *Resilient Agriculture* offer us glimpses of this resilient food future. The producers featured in these books blend traditional wisdom, the latest science and technology, and their own lived experience to tend unique agroecosystems shaped by the barriers and opportunities of place. These resilient producers do much more than simply grow high-quality food using farming practices that have restored the health of their land. They all reach beyond the farm and into the food system to support sustainable foodways in the places they call home.

All of them produce on-farm many of the resources they need to farm and manage their farm landscapes in ways that promote the ecosystem services needed to regenerate those resources. Many have diversified their product mix by producing processed foods on the farm. Nearly all of them market a significant proportion of their products directly to local consumers or into cooperative regional wholesale networks.

All of these farmers and ranchers have benefited from and contributed to the work of a small but committed community that has worked together in communities across our nation to imagine and produce sustainable food. Together they have championed a new American foodway—a way of feeding ourselves that invites us all, no matter where we stand in the food system, to make choices that heal the damages done by industrial food and promote ecological, social and economic well-being of land, people and community.

What will it take to make the shift to this new American foodway? Simply put, in resilience thinking terms, we must transform the operating context of American food and farming. We can begin by taking advantage of the lessons learned in the contemporary sustainable food movement. A movement that—although incomplete and imperfect—has devoted considerable effort to understanding what it means to feed ourselves without harm to land, people and community.

From Land to Mouth: In Search of Sustainable Food

In the latter half of the twentieth century, a growing awareness of the environmental, social and economic harms of industrial food systems drove a search for solutions that emerged as the sustainable agriculture movement. Defined by congress in 1990 and supported by a new federal program—the Sustainable Agriculture Research and Education program—much of the early investment in sustainable agriculture was focused on collaborative, on-farm research and development devoted to addressing regional farming needs. As the movement gained momentum, attention shifted to reimagining the food system in response to the increasingly globalized, concentrated and corporate-controlled food system that presented formidable barriers to the widespread adoption of sustainable agriculture. The sustainable food movement created a space to examine the sustainability challenges of industrial agriculture. A space that eventually came to include the whole of the global industrial food system—from land to mouth—in the search for sustainable solutions. A search that would eventually come to draw on the wisdom of the many cultures who know how to live well within the ecological limits of the places they call home.

Indigenous Foodways

For millions of years, *Homo sapiens* worked in community to feed on the plants and animals that lived in the lands they called home. Satisfying this basic need to eat drove the evolution of a wide diversity of foraging strategies shaped largely by local ecological conditions. In diverse ecosystems across the Earth, humans, like all animals, were nourished as part of the native food web. Archeological evidence suggests that by the end of the last ice age, about twenty thousand years ago, humans had evolved a good life based on foraging. Food foraging sustained stable, healthy human populations for millennia, but about ten thousand years ago, in different places around the world, the human population began to grow. Wherever that happened, people began to change the way they ate.

At different times within the Neolithic Period (10,000–2,000 BCE) and at different places around the world, foraging cultures slowly evolved one of three distinctly new foodways that included the careful cultivation of specifically selected and improved plant and animal species.[4]

In the desert grassland biomes of the world, where conditions are too dry for the cultivation of plant foods, people first hunted grazing animals and then domesticated some—sheep, goats, cattle and camels—to sustain themselves with a type of foodway that anthropologists have named pastoralism. In wetter forest biomes, horticulturalists began to cultivate the wild plants and animals that they depended on for food in forest clearings that they rotated through the landscapes of their home. On the fertile floodplains of the great river valleys in the world's savanna biomes, sedentary agriculturalists began to cultivate wild grains and grazing animals in permanent settlements. Even though these new foodways evolved in very different ecological circumstances, they share a number of characteristics that offer important examples of how to feed ourselves within the ecological limits of place.

First, all are embedded in local ecosystems and depend on healthy ecological processes to provide water, nutrients, pest suppression, waste disposal and other services needed to deliberately manage plants and animals for food. In every essential way, these early systems of agriculture were well-adapted to local ecological resource limits.

Second, all of these foodways took advantage of the many benefits associated with keeping livestock. Most importantly, these strategies recognized that animals are an efficient strategy for producing food from plants that are inedible to humans, for gathering and storing foods, and for utilizing food wastes. The ruminant animals—sheep, goats, cattle, camels—convert inedible grasses and forbs into high-quality meat, milk and eggs for human consumption, while others—dogs, swine, poultry—

can consume a wide variety of foods and serve as garbage collectors and recyclers. Animals also function as efficient "biological silos" for storing excess production for later consumption. Ruminants are particularly useful in this regard because they can survive for long periods without food.

Third, along with food, these Indigenous foodways produced another important product: an energy profit. Each produced more food energy than the energy expended in production, although none came close to the energy profit estimated for foraging of about 40 calories for every labor calorie invested. Pastoral and horticultural foodways yield about 11 calories, and sedentary foodways about eight for every calorie of labor (mostly human) invested. In comparison, industrial agriculture flips this energy relationship on its head. Instead of an energy profit, industrial agriculture produces an energy deficit that ranges from about seven calories invested for every vegetable calorie produced to about 32 calories invested (mostly as fossil fuel energy) for every calorie of beef produced. As a whole, it has been estimated that the U.S. food system requires about seven calories of energy to produce one calorie of food.

Ultimately, the problem of feeding our species boils down to a question of how best to manage five basic agricultural resources: land, water, energy, animals (including people) and plants. Pastoralists focused on managing livestock because grasslands have few native edible plants and shifting, uncertain rainfall. However, grasslands have a lot of grass and mobile grazing animals whose milk, blood and meat provide sustenance to humans. Horticulturalists did not have the benefit of large domesticable animals but had the advantage of large expanses of biodiverse forests and plentiful rainfall, and so they focused on cultivating food forests and woodland animals. Sedentary agriculturalists had the advantage of a diversity of edible grasses and other edible plants, regular flooding to regenerate crop nutrients in the soil, and large domesticable animals and so they developed agricultural systems that depended on both.

Thinking about the lessons these early foodways teach us raises some questions about how we might reimagine the global industrial food system that feeds us today. What if we could figure out a way to feed ourselves that recognizes and takes advantage of regional ecological limits? What if we could figure out a way to feed ourselves without the need to import nutrients, water and pesticides or export wastes? What if our foodways could again produce an energy profit along with nutrient-dense, culturally vibrant whole foods for all? What if our foodways could again sustain the health of land, people and community through relationships of respect and mutual benefit? What if our foodways could enhance our commonweal? And what if our foodways could reverse climate change? These are the kinds of questions raised by those interested in exploring the sustainability of the U.S. food system.

The Good Food Movement

From its earliest days, the contemporary "good food" movement has looked to sustainability principles as a way to address the growing environmental, social and economic issues created by the global industrial transformation of the U.S. food system in the latter part of the twentieth century. Sustainability inspired a generation of leaders to recognize the ecological impacts of industrialism. The power of sustainability as a driver of change is beyond dispute: within ten years of the Bruntland Commission's definition of sustainable development,[5] environmental sustainability principles were in use by government, business and the public throughout the world.

This cultural excitement around the idea of sustainability helped to shape the earliest expressions of the contemporary sustainable food movement in the U.S., which championed broader thinking about both the intention and the effects of the new relationships cultivated by global industrial food.[6] This work explored the potential to return to food system relationships that promote commensal community,[7] identified the growing distance between producers and consumers as a key sustainability concern,[8] and identified local food as the solution.[9] A multitude of individuals and organizations began to approach the question of sustainable food, each bringing valuable perspectives that enhanced our collective understanding of what it takes to sustain food and farming over the long term.

In the last decade, these different strands of thought have begun to converge in a powerful new vision of the future of food. A future that is equitable, sustainable, regenerative and resilient. It is a vision that is well within our grasp. It is a vision that—just like resilient agriculture—is already on the ground and growing, tended by people secure in their belief that we can find better ways to feed ourselves in community.

The stories shared by the farmers and ranchers featured in *Resilient Agriculture* show us how to cultivate this future, from the production of foods like fruit jams from Ela Family Farms, cooking oils from Quinn Farm and Ranch, Shepherd Farms pecans, drinking vinegars and hard ciders from Almar Orchards and Cidery, Rid-All's healthy green elixirs, humus from Zenner Farms, and White Oak Pastures meats, to collaborative production, processing and distribution networks like Shepherd's Grain in the Northwest, Hickory Nut Gap Meats in the Southeast and Strauss Dairy in California.

These producers, and many others just like them, are the people who supply your local cooperative grocery store and your local food restaurants. They are the people you see at your local farmers market, or at your community supported agriculture pickup, or at the community garden in your town. They are the people who gather at sustainable food and farming conferences every year throughout the country to share what they know about the art, science and spirit of growing, processing, preserving, preparing and celebrating locally grown foods in community. They have nourished and been nourished by decades of sustainable food activism, tirelessly working to transform the global industrial food system. Although a global movement informed by many, the stories of three movements that have come together in recent years holds particular significance for U.S. food resilience.

The Community Food Security Coalition formed out of the sustainable agriculture movement to promote community-based capacity for local food production, processing and marketing as a solution to the twin challenges of farm profitability and food insecurity. The One Health movement explores the public health effects of industrial food. The Food Justice movement works to address the structural race, class and gender inequities perpetuated by global industrial foodways. In recent years, these three movements have come together with the growing realization that it will take nothing less than the decolonization[10] of industrial food to realize the promise of sustainable foodways.

Cultivating Community Foodways

The Community Food Security Coalition was a diverse alliance of over 500 organizations in the U.S. dedicated to cultivating strong, sustainable, local and regional food systems to ensure access to affordable, nutritious and culturally appropriate food to all people at all times. Using a sophisticated blend of training, networking, and advocacy centered on local food projects, the alliance served as a kind of national mutual aid collective that has been credited with the emergence of the "good food" movements in the early years of this century.[11] For almost two decades, the alliance supported innovative projects and programs designed to connect eaters to the land and to food through farm-to-school and community gardening programs, farmers markets, new farmer projects and community supported agriculture. Although founded with a goal of transformative food system change, over time, market-based, local food systems strategies came to dominate the work of the alliance.

As our understanding of community food systems matured, some of the sustainability benefits assumed by the promoters of local food came into question.[12] New research exploring the popular concept of "food miles" as a meaningful measure of food sustainability challenged assumptions about the environmental benefits of local food.[13] When Walmart, the largest grocer in the United States, announced it would begin selling local food, some suggested that local food would suffer the same loss of integrity as organic food with the rise of "industrial organic."[14] Others offered evidence that the social benefits associated with local food networks may not survive scaling up because increasing the distance between producer and consumer erodes the social capital cultivated by direct markets.[15]

As the community food movement continued to explore these considerations of scale and integrity, the health professions began to grapple more publicly with some fundamental questions about the growing harms of industrial food on the health of land, people, community and the planet.

From Sustainable Diet to Planetary Health

Dietitians were among the first in the human health professions to explore sustainability. In a 1983 address to the American Dietetic Association, Kate Clancy proposed a "sustainable diet" to provoke discussion within the greater medical community about the impacts of industrial food. Clancy's use of sustainability as a framework for dietary guidance integrated multiple strands of thought—some old, some new—about the relationships between human and environmental health. Twelve years later, Joan Gussow was proud to note the nutrition profession's contribution to what would become an enduring idea: the choices we make about the way we eat shape our world.[16]

During the first decade of this century, thinking about human health in the context of food systems attracted the increasing interest of other groups in the health and planning professions. Veterinarians raised concerns about the widespread use of antibiotics in industrial meat production, while public health officials and land use planners began to explore how the operating context of the U.S. food system promoted some of the most critical health issues of our times. Making the connection between how what we eat shapes the health of individuals and communities, nutritionists continued to press their own profession to recognize all the ways that the global industrial food system creates barriers to healthy food choices for all.

This systems approach to health was formally articulated in an unusual joint statement calling for food system transformation in 2010 by the American Dietetic Association, American Medical Association, American Nurses Association, American Public Health Association and the American Planning Association.[17] The statement defined a sustainable community food system as one that "integrates food production, processing, distribution and consumption to enhance the environmental, economic, social and nutritional health of a particular place." The statement made explicit reference to "the interdependent and inseparable relationships" within the food system that create sustainable behaviors such as health, diversity, equity and resilience.

These efforts to connect the dots between food and health have matured into the planetary health movement which seeks to understand the human health impacts of industrial disruptions of Earth's natural systems.[18] Among the challenges addressed by this new movement are the human and planetary health consequences of global-scale changes such as biodiversity loss, environmental pollution, urbanization and climate change. Planetary health workers seek to understand how, for example, the loss of biodiversity and continued human encroachment on wild landscapes create the conditions for new diseases such as COVID.

The planetary health movement confronts three kinds of twenty-first century challenges to human health:

- *imagination challenges* such as the failure to account for long-term social-ecological consequences of human progress
- *research and information challenges* such as the failure to promote holistic and regenerative health strategies
- *governance challenges* such as delayed environmental action by governing bodies because of unwillingness, uncertainty or inability to cooperate

The movement promotes collaborative research across multiple sectors of the economy that are critical to human health and well-being, such as energy, agriculture and water. The goal of this work is to provide the science-based information that policymakers need to develop and implement holistic solutions to the impacts of global environmental change.

The Planetary Health Alliance is a diverse global consortium of over 200 organizations that are committed to advancing planetary health research, education and policy. Since its launch in 2016, the Alliance has supported a wide range of sustainable foodways projects, including the development of sustainable food education programs for health professionals, the release of the world's first planetary health diet in 2019,[19] and C40's Good Food City Declaration which commits signatories to reduce carbon emissions and increase resilience of city food systems by working with citizens to adopt the Planetary Health diet by 2030.[20]

From Food Security to Food Justice

Although social justice informed sustainable foodways thinking from the earliest days of the contemporary sustainable agriculture movement,[21] critics both within and outside of the community food movement have long urged a deeper recognition of the inequities perpetuated by industrial foodways. These voices encouraged the community food movement to put more emphasis on the well-being of all the people that work to feed us—particularly farm and food system workers—in rural communities and communities of color. These harms of the global industrial food system are rooted in historical patterns of access to and exclusion from resources based on race, class and gender that shape our relationships to land, people and community to this day.[22]

These voices urge us to acknowledge that the global industrial food system rests on a foundation of colonial thinking that justified genocide, theft, rape, murder, slavery, racism and oppression to support the good life of an elite class of Europeans. This legacy of colonialism is kept alive this day by our willingness to eat from a food system that requires farm families to subsidize farm income with off-farm employment, depends on people of color to labor in working conditions that are not tolerated by white U.S. workers and perpetuates the fragility of our nation by fueling agricultural injustice, food insecurity, degenerative disease and climate change.

Who is included in community, where does it begin and end, and who does the community food movement serve? These questions invite us all to reflect on both the intent and the effect of a movement that has been dominated by progressive, white, middle-class leaders who have failed to recognize the inadequacies of a change strategy focused primarily on market-based solutions to the harms of industrial food.

Market-based strategies have yielded some sustainability benefits, such as increased food literacy and direct market opportunities for producers and more diversified food purchasing options for recipients of federal, state and community-based nutritional assistance programs. As demonstrated during the early days of the COVID pandemic, these strategies have cultivated local and regional food networks that increase the functional and response diversity of our national food system. These benefits are welcome and important, but upon closer examination, it is clear that the primary beneficiaries of nearly half a century of sustainable food activism are progressive white middle- and upper-class eaters and the businesses that cater to them. Although the intent of the good food movement may have been food system transformation, the effect has been to protect the existing industrial system through economic development that tends to reinforce existing race, class and gender inequities.

This critique of the sustainable food movement has been recognized in the U.S. as a call for "food justice" through the different, but related, food sovereignty, food democracy, food solidarity and fair trade movements. Integrating lessons learned in the social justice and environmental justice movements, the food justice movement supports work led by Indigenous peoples and people of color to confront the structural inequities in the food system. This work emphasizes food system relationships—those that harm, those that heal and those that have the power to cultivate cooperation, trust and sharing economies that cultivate the health and well-being of land, people and community.

The Growing Food and Justice for All Initiative (GFJI) is just one example of the many organizations working towards the just transformation of community foodways. Launched in 2008, the Initiative promotes food justice in the tradition of Martin Luther King Jr.'s "Beloved Community." Working to dismantle racism and empower low-income communities and communities of color through sustainable and local agriculture, the GFJI shifts the role of immigrants, Indigenous peoples and other communities of color from laborers to entrepreneurs.

According to Erica Allen, co-founder of the GFJI,[23] the goal of the Initiative is to empower and challenge people to do the work of removing the obstacles of racism and the other "isms" that stand between all people and a fair and equitable food system. GFJI members lead research, policy-making and projects that advance the organization's antiracist and economic objectives through community formation, political activism and the identification of effective strategies for leveraging food-based economic development. The members of GFJI and other food justice organizations invite us all to reconsider the contradictions in our own ways of thinking about food—history, core concepts, values and practices—in order to find new ground for collaborative actions that disrupt the colonial roots of global industrial foodways. Over the last decade, the Initiative has nurtured the development of new food justice organizations led by Blacks, Indigenous people and people of color as well as continuing to work with other community food organizations to more fully integrate antiracist values and goals into their work.

One measure of the success of the food justice movement is the growing recognition within U.S. sustainable agriculture and food organizations of the need to promote racial healing and anti-racism practices within their own organizations. Amplified by the many inequities in the U.S. food system revealed by the COVID pandemic,[24] this new awareness prompted the National Sustainable Agriculture Coalition to declare, "There is no future for sustainable food or farming without racial justice."[25]

13

Adding Resilience to the Menu

A NEW AWARENESS OF climate change, coupled with a series of global economic shocks in 2008, 2012 and most recently in 2020, have driven increased interest in the resilience benefits of sustainable foodways over the last decade. Business and government leaders, NGOs and the media increasingly seek food system solutions to a multitude of global challenges.[1] The solutions that receive the most attention are typically those that protect business as usual in the food system. They share a number of common characteristics.

Some of the most popular solutions depend on new industrial technology, for example, synthetic food and farming, lab-grown proteins, vertical agriculture and big data applications of all kinds. Other popular solutions focus on changing individual behavior, like eating more plants, or wasting less food or paying farmers to adopt new practices that capture and store carbon on their farms. There has been an explosion of new "climate friendly" diets for us to try, including the ethical omnivore, flexitarian, vegan, vegetarian and planetary health to name just a few.

Another prominent theme of food system solutions is to focus on just one sector (production, processing, marketing or consumption) and one dimension (environmental, social or economic) within that sector. For example, some solutions address production sector issues such as the loss of farmland, exploitation of farm labor or fertilizer and pesticide use. Processing sector solutions might highlight worker safety, the dangers of concentration and consolidation or the public health harms of processed foods. Distribution/marketing sector solutions often advocate for fair wages, food security, healthy choices and the reduction of food waste.

The problem is, these popular food system solutions typically require little change in the identity, structure or function of global industrial foodways. What these solutions do best is protect the existing system, even though they may be promoted by organizations that seek food system transformation.[2] And so here we are, right back to the problem of the six blind men who can't make sense of the elephant. Unable to comprehend the whole of the food system, we are caught in a dangerous and ultimately futile exercise of tweaking the existing system in order to keep it functioning despite its well-documented fragility.

We don't have to accept these band-aid solutions. At best, they buy us a little more time to rest in a comfortable state of denial. At worst, they create new harms. Deep down, we all understand that these kinds of solutions—actions that focus on symptoms rather than causes—don't work. We've known this for a long time. Our desire for silver bullet solutions is an outright refusal to connect the dots between symptom and cause, because to do so—to put the focus on relationships, on the structure, function and purpose of systems—is a direct threat to business as usual. Once you understand this, you will see this kind of thinking nearly everywhere, used by most everyone. We can't help it—this failure to see the whole is the defining legacy of our colonial past. It is a way of being in the world that shapes our present reality. It is a legacy that limits the way we think about the possibilities of our future.

There is an alternative. A completely new kind of solution is emerging in the sustainable food movement. A solution that integrates the lessons learned over the past 50 years—lessons about local food, sustainable diets and food justice—to create a radically new perspective on the future of food. It is a solution that invites us to recognize that it is the quality and kind of food system relationships that determine the well-being of land, people and community.

This new solution solves the problem of the six blind men by inviting the whole village into the conversation about food solutions. This solution engages the light in our nature by igniting those biologically-wired behaviors that help us live well in community: honesty, equity, responsibility, compassion, respect and courage. This way of thinking calls on our innate capacity to be generous, to love and care for each other. It takes advantage of what we know we can accomplish when we listen to different perspectives with respect, create shared a vision of the future and then work together in community to achieve it. This new kind of thinking points to the region as the right size for food system resilience.[3]

Just a little over a decade ago, researchers working in the Collaborative Initiatives program at the Massachusetts Institute of Technology sought to discover solutions to the public health harms of the global industrial food system.[4] Using design thinking, this group identified the current structure and function of the U.S. food system—the way food is produced, processed and distributed—as a primary culprit in America's obesity epidemic and high rates of chronic disease. They recommended "foodshed reform" as a critical step to overcoming national challenges in public health, environmental quality, energy use and economic inequity. The Initiative's researchers proposed to solve these challenges by building on existing local and regional foodsheds to transform the U.S. food system into a nationally-integrated system of regional foodsheds. A follow-up study piloted a GIS-based scenario analysis tool to guide investments in such food system re-regionalization.[5]

The foodshed concept has proven to be one useful framework for exploring food system sustainability.[6] Modeled on the watershed, the concept of the foodshed offers a way to visualize the geographic area through which food flows from producers to consuming communities. Along with the shift to the regional scale, the Collaborative Initiatives researchers identified several additional changes that would be critical to foodshed reform: a transition to more diversified production systems; the development of a regionally-based processing and distribution infrastructure; and new models of food retail. Ironically, these recommendations echo those of New York City planners concerned about the impact of industrialization on the city's food self-reliance at the turn of the twentieth century.[7]

The Regional Roots of Resilience

The ideal regional food system will be one that produces multiple social, economic and environmental benefits to all the region's inhabitants while contributing significantly to their food needs.[8] Much of the production, processing, distribution and consumption will take place within the region and will be achieved through complementary relationships within and between a diversity of scales from the local to the global. Cultivating these kinds of regional food systems offer unique advantages to enhance food resilience at both the local and the national scale. Ideal regional food systems are self-reliant, not self-sufficient—in other words, they are capable of providing for regional food needs through local and regional production supplemented by limited and mutually beneficial trade relationships with other regions.

The resilience of the regional scale is generated by a network of place-based foodways relationships between land, people and community that cannot be created at larger (national/global) or smaller (local) scales. Resilient regional foodways are not simply all the local food systems in a region, nor are they a scaled-up version of local food. Food system resilience at the regional scale emerges through the complex network of complementary relationships within and between local and regional food systems and the national food system. The cooperative relationships and resources in regional food systems are unique to the regional scale and offer many advantages that contribute to foodways resilience.

The expanded land base of a regional, compared to a local, food system offers sustainability and resilience benefits along four dimensions—food supply, natural resource management, diversity and economic development.[9] Historically, regional food systems satisfied a significant proportion of local food demand, and recent research suggests that many regions in the U.S. could still do so.[10] Regional approaches to the management of the natural resources upon which food production depends—particularly land and water resources—have a long history of proven effectiveness. Regional populations have unique opportunities to influence both local and national policies and programs that shape the operating context of the global industrial food system.

Regional landscapes have the ecological diversity needed to cultivate the agroecosystem behaviors that enhance self-reliance such as the internal sourcing of critical production inputs and marketing of products. Regional ecological features can contribute to a shared sense of identity within a population—for example, the Chesapeake Bay, the Southern Appalachians, the Driftless Region or the Sacramento Valley—that can help to encourage collaborative action through our innate awareness of the sounds, sights, tastes, textures and seasons of the landscapes we call home.

A regional perspective is a useful position from which to explore inter- and intraregional relationships that determine the patterns of trade, development, wealth, education, population, transportation and waste management that together shape food system sustainability and resilience. The economic benefits associated with shorter supply chains can still be realized at the regional scale, but greater efficiencies in food storage, processing and distribution can be gained through the cooperative development of regional foodways infrastructure. Finally, a regional focus links urban markets and rural production areas, offering an opportunity to include intraregional interactions such as trade, development, transportation networks and other system elements that will become more important as urban areas continue to increase in significance in coming years.[11]

Food system-focused assessment and planning as a path to sustainable regional economic development has increased in the last decade. Most of these projects have focused on enhancing food system equity by overcoming market-based barriers to the development of regional food economies. Although resilience is not a stated goal in much of this work, such projects have the potential to enhance community resilience because their findings tend to promote increased local and regional production, processing and marketing of sustainable food. Better integration of food and farming into climate resilience planning would dramatically increase the potential of regional food systems to promote climate justice and food justice goals while enhancing community climate resilience.

Some notable regional food system planning projects with explicit resilience goals are described in Table 13.1. Resilience was simply a high-level goal in many, but a few of these projects integrated resilience thinking more deeply, especially those conducted in Iowa, Vermont, Philadelphia and New England. Only one connects the dots between land, people and community: The New England Food Vision.

The New England Food Vision was a uniquely collaborative project led by Food Solutions New England to explore the impact of dietary choices on sustainable food production in the region.[21] The sustainability impacts of three distinct diets—current industrial, USDA-recommended, and self-reliant—were investigated in terms of ecological resource concerns relevant to land, energy and water resources, climate change and biodiversity, and social resource concerns associated with cultivating regional food system capacity. Through widespread changes in the existing regional food system, including some controversial recommendations like the conversion of forests to farmland, the work found that New England could supply about 50 percent of its food needs by 2060, while setting aside adequate room for smart growth and keeping 70 percent of the region in forests.

An important finding of this work was that a substantial volume of food trade with other regions would be necessary to obtain the quantity and diversity of foods needed to meet the dietary requirements of New England's population. A companion report identified the policy changes that would be needed to support expanded production, strengthen food supply chains and enhance multi-state cooperation to achieve the vision of a more robust and resilient regional New England food system.[22]

In a paper[23] included in the Symposium on American Food Resilience,[24] Kathy Ruhf reflected on the development of the New England Food Vision. She suggested that the "soft working borders" of the New England states contributed to the success of the project, along with a legacy of trust and cooperation cultivated by decades of institutional collaboration. Kathy pointed out that although resilience has not been an expressed goal of many New England food initiatives, achieving

Table 13.1. Exploring Regional Food Solutions

Food system-focused assessment and planning as a path to sustainable regional economic development has increased in the last decade. This table presents some notable regional food system planning projects completed in the last decade that include explicit resilience goals. Resilience was simply a high-level goal in many, but a few of these projects integrated resilience thinking more deeply, especially those conducted in Iowa, Vermont, Philadelphia and New England. Only one connects the dots between land, people and community: The New England Food Vision.

Project	Area	Purpose	Outcome
Cultivating Resilience: An Iowa Food System Blueprint that Advances the Health of Iowans, Farms and Communities[12] 2011.	Statewide	Conduct a resilience assessment of Iowa's food system.	Assessment using 14 resilience indicators representing economics, environment, food and farm equity, and food access and health. Recommends investment of state economic development funds to create new local/regional food system infrastructure and support new small and mid-sized food production and processing businesses.
Regional Food System Plan for Vermont's Northeast Kingdom[13] 2011, 2016 Update.	3-county	Assess area food system using a soil-to-soil closed-loop model to support a vibrant food and farm economy through a regional planning process.	Food system assessment using 60 county level indicators for seven food system processes.
Vermont's Farm to Plate Strategic Plan[14] 2011, Annual Updates, Revised 2021.	Statewide	Develop and implement a strategic plan to achieve a food system that is economically successful, environmentally sound and socially just.	Annual assessment of 25 goals included in a ten-year strategic plan.
Eating Here: Greater Philadelphia's Food System Plan[15] 2011.	Parts of 5 states	Develop a plan to promote six core values of a sustainable and resilient food system in the Philadelphia Metropolitan Area.	Plan identifies opportunities to promote goals and associated assessment metrics.
A New England Food Vision[16] 2014, Update underway 2021.	6-state	Develop a practical vision of a regional food system that supports four core values: food rights, healthy eating, sustainability and community vitality.	Biocapacity footprint analysis of sustainability of three diet scenarios: current, USDA recommended and resilient. Recommends pathway to 50% regional food production by 2060 while supporting healthy food for all, sustainable farming and fishing, and thriving communities.

Table 13.1. (cont'd.) Exploring Regional Food Solutions

Project	Area	Purpose	Outcome
New England Feeding New England: Cultivating A Reliable Food Supply 2019.[17]	6-state	Ten-year initiative to advocate for state and regional planning to: (1) expand regional food supply chains and associated employment; (2) transition to climate-smart and climate-adaptive food systems; (3) promote local, equitable control of land and water resources for sustainable food production.	In 2021–22 will initiate: (1) coordinated and robust engagement with diverse food system stakeholders; (2) regional coordination of food system data and regional convenings; (3) strategic planning to reach goal of 30% regional food supply by 2030; (4) landscape analysis of climate-friendly production practices, emergency food planning and policy initiatives to promote resilient regional food supply chain.
Good Food for All Chesapeake Food System Assessment[18] 2020.	Parts of 7 states	Broad analysis of current barriers and opportunities to promote an equitable, sustainable and thriving regional food economy food in the Chesapeake Bay watershed.	Recommendations address critical gaps and challenges: attract new investments in regional infrastructure; engage schools and health care organizations; include social justice advocates; increase regional and statewide coordination.
Food for All Piedmont Triad Food System Assessment[19] 2021.	12-county	Develop a baseline for understanding the regional food system, examine economic opportunities for strategic investments, and create a sense of shared ownership and equity principles in order to enhance the equity and resilience of the regional food system in central North Carolina.	Reports innovative analysis of regional food geography. Recommendations in five focus areas to: enhance food justice, increase regional food production and processing, improve economic opportunity for farm and food businesses, enhance regional supply chains and leverage regional resources for food system change.
Greater Capital Region Food System Assessment[20] 2021.	11-county	Identify opportunities to enhance equity and economic resilience for regional food producers and low-income consumers in the 11-county regional foodshed supplying Albany, NY.	Recommendations for economic development strategies in four sectors (consumption, distribution, processing and production) to achieve three goals: provide low-income residents with consistent access to healthy and culturally important foods; provide farmers with equitable access to land, financing and profitable markets; and provide food businesses with high quality regional produce.

sustainable development goals with food system change was implicitly understood to promote regional resilience. She concluded that the region's physical and cultural geography contribute significantly to the potential for a self-reliant food economy, along with other important factors that help to shape this potential such as the high diversity in agricultural production systems, a limited land base relative to population size and a low level of dependence on federal farm programs and global trade.

As disturbances and shocks to the global food system from climate change grow more frequent and intense, metropolitan regions (areas with a concentrated population of 50,000 or more) have increased their interest in enhancing the resilience of their food supply. Much of this work focuses on options for managing short-term disruptions to food sourcing and distribution without any consideration of broader food resilience concerns. For example, New York City has an initiative to invest in making the city's fresh food distribution center at Hunt's Point more resilient to power outages, coastal flooding, job losses and other disruptions from extreme weather events. Although such planning is critical to disaster response, it is not designed to address the broader issues underlying the fragility of the city's food supply.

Metropolitan Foodsheds

The "city region" is an emerging, landscape-based sustainable development strategy that is well-aligned with regional food system planning for resilience. City region food systems are defined as the complex network of relationships that link urban, peri-urban (urban-rural transition zones) and rural areas within a geographic region through the production, processing, marketing, consumption and disposal of food. This new way of thinking about a city's food supply supports resilient sustainable development because it focuses on cultivating healthy regional flows of people, goods and ecosystem services.

Sustainable city region food systems can serve a just transformation of the food system by strengthening social relations, supporting participatory governance, reducing dependence on distant supply sources and cultivating healthy regional ecosystems.[25] Although still relatively new, there is growing evidence that the city region as a planning concept provides an effective strategy for food system change through sustainability-focused, multilevel governance initiatives. City region food system initiatives have increased food and nutrition security in vulnerable communities while enhancing the livelihoods of small farms and food businesses. By promoting the adoption of sustainable agriculture practices, city region food system development has restored degraded land and generated climate change mitigation and adaptation benefits. This new way of thinking about metropolitan food systems has also been proven to promote regional economic development, catalyze innovation and enhance public health.[26]

Despite the many documented benefits of city region food systems, metropolitan areas present some unique benefits and challenges to the development of sustainable and resilient regional food systems.[27] On the plus side, sustainable producers located in or near these areas have easy access to a large population of potential customers, high-value direct markets and value-added processing opportunities. Physical infrastructure for power, water, transportation and other resources is usually well developed, and proximity to the metropolitan core offers opportunities for meaningful off-farm employment for non-farming family members, as well as employment opportunities on farms and in the food system for the urban underemployed.

Metropolitan areas and the peri-urban periphery also present some challenges to food production. Land values are high, non-farming residents may object to farming operations, and access to traditional farm services may be difficult. Generally, urban supply chains are scaled for much larger wholesale volumes than can be supplied from farms within metro regions. Concentration in the grocery sector and a dearth of independent transportation partners are challenges for mid-scale farmers producing wholesale products for metro markets.[28] Traffic congestion around cities creates a costly barrier to efficient freight delivery of any kind and food is no exception. The existing relationships between businesses in food supply chains are predominantly national in scope, presenting a barrier to the development of regional connections.[29]

Competition with municipal and industrial uses for resources may also present considerable challenges, particularly as climate change intensifies, but close proximity of municipal, industrial and agricultural systems will also encourage innovations that increase the efficiency of resource use through sharing and recycling.[30] Ultimately, addressing the power dynamic between urban and rural areas will be key to reversing the extractive behavior of cities and putting into place systemic solutions that cultivate sustainable and resilient metropolitan regions.[31]

In recent years, a growing awareness of the sustainable development benefits of city region foodways has increased interest in sustainable food as an urban resilience strategy.[32] This new interest can be clearly seen in international networks that support local urban development efforts such as Local Governments for Sustainability[33] and the Milan Urban Food Policy Pact.[34] City region food projects carried out by members of these networks confirm the value of food-systems thinking to urban sustainability and resilience planning. Some promising focus areas for future food policy work by network members include engaging small and mid-sized cities, developing new integrative approaches to the urban-rural spectrum, exploring options for deep adaptation to climate change and promoting social innovation. More generally, because food connects all of the UN Sustainable Development Goals, network members find additional value in using food-systems thinking to integrate many SDG aims and priorities.[35] The city region food systems concept offers a promising new approach that integrates food systems thinking into visioning sustainable and resilient futures for land, people and community in metropolitan regions.

Planners in the U.S. have already begun to explore options for enhancing the sustainability and resilience of the ten U.S. mega-metropolitan regions which are projected to be home to more than 70 percent of the U.S. population by 2050.[36] These mega-metropolitan regions, linked by economic and transportation systems, history, culture and natural resources, are viewed by regional planners as an important opportunity to reimagine twenty-first century housing and urban development, transportation and water systems and to integrate into their work, for the first time, food and farming systems. This new way of thinking about the role of food and farming in regional sustainable development supports a powerful vision of a New American Foodways: a nationally integrated network of metropolitan foodsheds (See Figure 13.1).

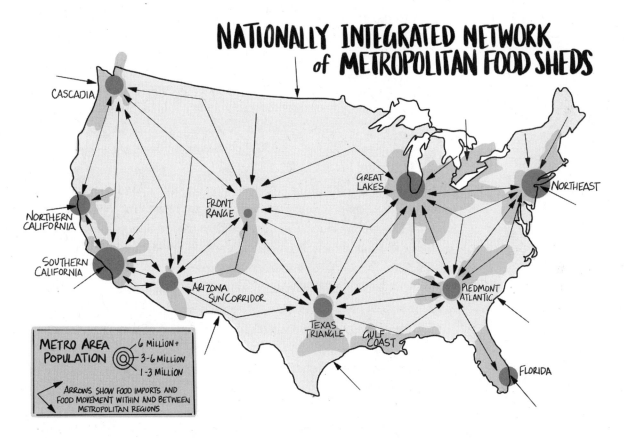

Figure 13.1. The New American Foodway? Regional planners have begun to explore the potential sustainable development benefits of incorporating food system planning into regional planning underway for ten U.S. mega-metropolitan regions. A rediversification and re-regionalization of the U.S. food system has been recommended as a critical step to overcoming national challenges in public health, environmental quality, energy use and economic inequity[37] and is well-aligned with the qualities and behaviors of a resilient national food system.[38] Credit: Caryn Hanna.

14

The Way Forward

THERE IS A GROWING SENSE in government, business and civil society worldwide that business as usual is no longer an option when it comes to food. It is widely acknowledged that the global industrial food system fuels the wicked challenges of our times—concentration of wealth, loss of biodiversity, energy use, population growth and climate change—that have put us on a path to planetary collapse. At the same time, more people are waking up to the potential of food and farming as a powerful climate solution. The question is not *if*, but *how* we change the way we eat. So many different solutions are bubbling up all around us—how do we choose among them?

Social-ecological resilience thinking offers new concepts, a new language and an effective framework for decision making that is uniquely suited to the novel uncertainties of our times. Resilience thinking can help us identify solutions that cultivate the kinds of social-ecological relationships that produce high response, recovery and transformation capacity: networks of equitable relationships; regional self-reliance; and the local accumulation of community-based wealth, including natural, human, social, financial and technological resources. And resilience thinking helps us to name and then work to heal the harms of industrial thinking that continues, to this day, to cultivate the fragility of our land, people and community.

It's not by chance that the core principles and practices of the U.S.

sustainable agriculture and food movements cultivate resilience, because both sustainability and resilience thinking are contemporary expressions of the wisdom known to our Indigenous ancestors about how to live well in community. Although industrial thinking encourages us to deny, denigrate and ignore this wisdom, it lives on as points of light within the darkness of our colonial legacy to illuminate the path to our resilient future. I can see this light clearly in the early organic farming movement, in biodynamic agriculture, in mutual aid societies, in the labor movement and the cooperative movement, and in the wave upon wave of social justice movements that have confronted the harms of industrial thinking for centuries.

The sustainable food and farming movement is one point of light, however incomplete and imperfect, that has tended to this wisdom and worked to keep it alive. Sustainable food and farming activists have prepared the ground, sown the seeds, harvested and shared the bounty of the Earth in a celebration of all that we know about how to live well on this planet. We have worked together, in community, to find new ground and plant it with the seeds of our resilient future.

How do we get to that future from here? Resilience thinking can help.

To get started, we can use the three rules of resilience to answer questions like:

- Does the proposed solution promote a diverse network of reciprocal relationships? Resilience thinkers value solutions that enhance mutually beneficial foodways relationships that cultivate response, recovery and transformation capacity.
- Does the proposed solution promote regional self-reliance? Resilience thinkers value solutions that reduce foodways dependence on the import of critical resources and the export of products and wastes.
- Does the proposed solution promote the local accumulation of community-based wealth? Resilience thinkers value foodways that generate a diverse, locally-controlled portfolio of high-quality resources required for community well-being.

Next, we can dig a little deeper by exploring the potential of the proposed solution to promote the resilient foodways behaviors shown in Figure 14.1 and described in Table 14.1 to answer questions like:

- What are the key strengths of the solution? Resilience thinkers value solutions that directly address the root causes of fragility.
- What are the key weaknesses of the solution? Resilience thinkers value solutions that serve the commonweal.
- What key external threats are addressed by the solution? Resilience thinkers value solutions that offer both specified and general resilience benefits.

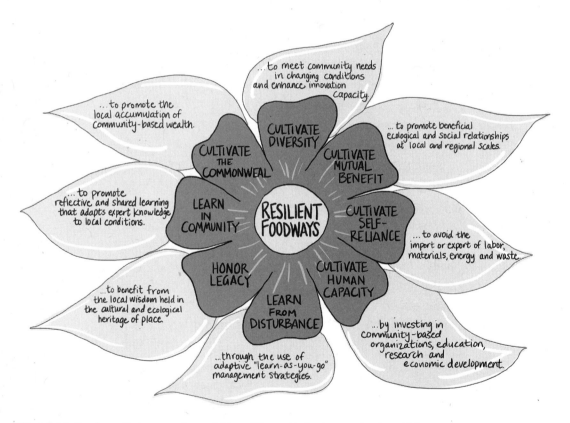

Figure 14.1. Foodways that express these eight qualities and behaviors tend to have high response, recovery and transformation capacity.[1] Credit: Caryn Hanna.

- What key external opportunities are addressed by the solution? Resilience thinkers value solutions that promote the ability to respond to change in ways that avoid or reduce damage from disturbance, to recover swiftly and at low cost when there is damage, and to find opportunity in change.

Finally, we can examine the proposed food system to answer questions like:

- What is the management intent and effect of the solution? Resilience thinkers value solutions that clearly describe why the action is recommended and the likely effect of the action on foodway response, recovery and transformation capacity.
- How coherent is the management intent and effect of the solution? Resilience thinkers value solutions that are likely to have the intended effect.
- Does the solution add value to available adaptation options? Resilience thinkers value having access to a range of complementary solutions that can be used together in an adaptive strategy that protects, adapts and transforms the system.

Successful system transformations involve changes that shift the operating space to support community-based actions that cultivate resilience and penalize (or at least stop subsidizing!) actions that maintain or enhance the fragility of the system. Where will the most popular foodways solutions take us? Where do your own favored solutions lead? No one knows what the future holds, but resilience thinking can help us confidently make the shift towards a foodway that serves land, people, and community better than the way we eat today.

The resilient way forward requires a transformation, a change in form and function and, perhaps most importantly, a change in the purpose of the U.S. food system. It is not one path but many paths, not a silver bullet but silver buckshot, a diversity of individual adjustments and adaptations within industrial foodways that will move us from fragility to resilience.

Table 14.1. Eight Behaviors of Resilience in Regional Foodways

Resilience Behaviors	Examples of Resilient Foodways Practices
Cultivates Diversity	Diversified, cooperative food systems that promote ecological and social/cultural diversity. Production systems that integrate annual and perennial crops, livestock and wildlife; on-farm processing and waste recycling; multiple food distribution transportation modes and food storage locations; diverse retail outlets; public policy promoting food system diversity, food access, diverse ownership of food system businesses.
Promotes Mutual Benefits	Diversified production systems that promote ecosystem services such as pollination, soil health, wildlife habitat; cooperative supply chain relationships with multiple producers, processors, suppliers and markets; participation in local purchasing and marketing cooperatives; local ownership of food-supply-chain businesses and small business viability; shared supply chain risks, costs and benefits; democratic governance.
Cultivates Self-Reliance	Food system operations contribute to healthy regional energy flow, water and mineral cycles; promote plant, animal and human health and community well-being; require minimal import of critical materials such as energy, water, nutrients, raw materials and labor, and minimal export of raw materials, processed products and wastes.
Cultivates Human Capacity	People organize into grassroots networks and institutions such as co-ops, labor unions, farmers markets, food policy councils, community gardens and advisory networks; food system relationships are based on strong communication, trust, and shared risks and benefits; investments in infrastructure, institutions and governance supports community-based education, research and development of locally-owned food system businesses.
Learns from Disturbance	The capacity to cope with disturbances or shocks is promoted by the use of adaptive management strategies that use regular monitoring and formal evaluation of system performance to guide decision making.
Honors Legacy	Activities that explore, preserve, honor and celebrate the cultural and ecological heritage of place, such as maintaining heirloom seeds and heritage livestock, and documenting regional foodways.
Learning in Community	The use of adaptive management strategies that use regular monitoring and formal evaluation of system performance to guide decision making; participation in peer-to-peer and community-based education, research and development networks.
Cultivates the Commonweal	Profitable production, processing and distribution of food products; smooth intergenerational transfer of locally-owned farms and other businesses in the food system; the local accumulation of wealth and fair labor relations throughout the food system.

We can begin now to redirect private and public investment to create a New American Foodway, a nationally-integrated network of regional metropolitan foodsheds that support the well-being of land, people and community by cultivating high social and ecological response, recovery and transformation capacity. As a first step, we can learn from those cities and states that are already working to create our nation's most innovative foodways,[2] powered by a vision that recognizes the regional roots of resilience and embraces just transformation.

Resilience invites us to see the challenges of these times with new eyes, to ask different kinds of questions, to embrace uncertainty and to find opportunity in change. We must accept that we will not be able to burn or build our way to a climate change solution. We must accept that we cannot thrive on technology alone. We must make a fundamental shift in the way we relate to our past, examine our present and imagine our future.

As we consider this path, it's important to remember that it does not lead us back to the future. Since colonial times, American foodways have degraded land, people and community. It is also important to remember that resilience is not just another name for sustainability. Resilience changes the rules of the game. It changes the way we define problems and the way we evaluate solutions. Resilience thinking can help us step forward into something new, into a future that has the potential to be as different from our current reality as agriculture was to our foraging ancestors.

Twelve Things You Can Do
to Cultivate a Resilient Agriculture

1. Stop wasting food.

2. Stop buying silver-bullet brands.

3. Stop thinking about resilience as bouncing back.

4. Stop believing that somebody else will save us or that nobody can save us.

5. Learn a new resilient foodways skill.

6. Learn how your ancestors nourished their community.

7. Learn more about the land, people and communities that feed you.

8. Learn more about climate risk, resilience planning and climate justice in the place you call home.

9. Follow the three rules of resilience at home, in your profession and in your community.

10. Participate in your regional foodshed.

11. Participate in just transformation.

12. Remember that privilege is power and use yours for good.

PART 4

REAL WORLD RESILIENCE: STORIES OF LAND, PEOPLE AND COMMUNITY

You just don't know what's around the next corner,
so you have to prepare for the worst.
Hope for the best of course, but you know,
hope is not a plan.

— Gary Price, The 77 Ranch,
Blooming Grove, Texas

THE STORIES THAT I SHARE here represent a diversity of answers to one of the most important questions of our times: How do we feed ourselves in a world on fire? Some of these stories are continued from the first edition, but I have gathered most of the stories shared here in the past few years. These are brave stories, being lived by real people who have made the choice to cultivate a resilient food future, despite the many barriers created by global industrial food. Their stories offer us a glimpse into some of the many practical and proven ways that food and farming systems can be organized and managed to cultivate diverse networks of mutual benefit, regional self-reliance and the commonweal. The people who share these stories grow vegetables, fruits, nuts, grains, milk and meat in every part of our nation with the intention of producing healthy food along with a multitude of other social and ecological benefits.

Each producer in this volume brings a unique set of advantages and disadvantages to their work that has shaped and continues to shape their choices about why and how to grow food. Many find opportunity in their family legacy, while others struggle to overcome it. Some have just started growing, others have grown for decades and still others have a way of knowing the land they farm that stretches back sometimes four or five generations. Despite the many differences among them, these farmers and ranchers are united in purpose: they all want to grow food that promotes healthy land, people and community.

Each has a story to tell, a unique perspective to share about the challenges of growing food in a changing climate. Together, these stories weave a rich tapestry of place-based wisdom accumulated through lived experience—failures that teach, successes that inspire—that can help guide us through the many challenges on the path to a New American Foodways:

- Hannah Breckbill purchased her land through a community co-operative.
- Keymah Durden worked with childhood friends to rebuild their community by plugging into ancestral wisdom via the "soul of soil."
- Ira Wallace and Mary Berry are excited to try transplanting vegetable seed crops using their new human-powered no-till transplanter imported from South Africa.
- Albert Straus finds opportunity in new technologies that transform manure into compost to improve the health of his pastures and renewable energy to power his farm.
- Mark Shepard selects for hyper-precocity in perennial nut crops as a risk management strategy.
- Rebecca Graff and Tom Ruggieri added vegetable ferments to their product mix to improve the health of their soils.
- Gail Fuller retreated from increased flooding by selling his fourth-generation family farm and relocating operations to higher ground.
- C. Bernard Obie is grateful for the improbable circumstances that brought him back home to become the fourth generation to farm his family's land.
- A.G. Kawamura grows fresh produce on rented ground in residential developments, military bases, city parks, schools and abandoned orange groves in Los Angeles.
- Pam Dawling has grown vegetables and fruits in community for more than 50 years and has yet to earn a dollar through sales.
- Fresh off the battlefield, Jordan Settlage was determined to find a way to farm that celebrated life.
- Bryce Lundberg appreciates the climate resilience benefits of his family's long commitment to the principle of "leaving the land better than you found it."

- Walker Miller added pine trees to his blueberry fields to take advantage of the "hand shake" between the mycorrhizae associated with the pine and the blueberry.
- Jamie Ager sees the growing public interest in "climate-solution-style" farming as a new opportunity for his regenerative meats business.

I see some common threads amidst the colorful tapestry woven by the diversity of stories shared here and in the first edition. Although many are not sure what they think about the cause, nearly all have stories to tell about weather that falls outside of personal and historical experiences of regional weather norms. Specific climate risks vary by region, specific location and type of production system, but producers everywhere tell of the challenges created by too much and not enough water, by too hot and too cold temperatures, often in the same season and sometimes in the same day. This "stacking" or accumulation of multiple climate disturbances within a growing season, both subtle and not so subtle, is becoming more common nationwide. These experiences of changing weather patterns are consistent with the climate science and offer us a real-world window into the consequences of our failure to act on climate change.

The greater challenges of farming within the limitations and amid the growing disturbances and shocks unrelated to weather are another common thread. Few of these farmers and ranchers say that the weather is the biggest challenge facing their operations today. Most point to social forces as the greatest risk to their sustainability: the challenges of operating within the continued industrialization and globalization of the food system, new one-size-fits-all environmental, health and labor regulations that do not recognize the relative risks and benefits of different kinds of production philosophies, and state and federal agricultural assistance programs that continue to privilege large-scale industrial production, processing and distribution systems through public investment in research and development, education, technical assistance, and the subsidy of critical irrigation, transportation and fossil fuel infrastructure.

These social forces serve to amplify climate risks to our nation because they degrade the resilience of an already fragile U.S. food system.

Another common thread is an appreciation for the climate resilience benefits of sustainable agriculture practices that enhance soil health, promote diversified operations and support high-value marketing. Most want to learn more about how the way they farm can help to slow climate change, many are investigating how to integrate agricultural climate solutions into their operations and some are leading the way. Based on their past experiences, most are upbeat about their ability to remain successful, no matter the weather, but many express concerns about the sustainability of their operations as weather variability and extremes grow more intense.

Chapter 15 shares the stories of sustainable producers using biodynamic, organic and regenerative farming philosophies to produce vegetables and other crops for local and regional direct and wholesale markets. These farmers manage operations ranging from 2 to 1,000 acres, on family farms, in intentional communities, on conserved land, community-owned land and urban land. Table 15.1 presents summary information about all the vegetable producer stories published in the first and second editions of *Resilient Agriculture* and where you can find them.

Chapter 16 shares the stories of sustainable producers using organic and climate-smart growing philosophies to produce fruits, nuts and other crops for local and regional direct and wholesale markets. These growers manage operations ranging from 22 to 106 acres on land that was restored with sustainable farming practices. Table 16.1 presents summary information about all the fruit and nut producer stories shared in *Resilient Agriculture* and where you can find them.

Chapter 17 shares the stories of sustainable producers using organic, climate-smart and regenerative growing philosophies to grow grains and other products for local and regional direct markets and national wholesale markets. These farmers manage operations ranging from 1,500 to 3,200 acres. Table 17.1 presents summary information about all the grain producer stories shared in *Resilient Agriculture* and where you can find them.

Chapter 18 shares the stories of sustainable producers using organic and regenerative farming philosophies to produce meat, poultry, pork and dairy products for local and regional direct and wholesale markets. These farmers manage operations ranging from 160 to 500 acres. Table 18.1 presents summary information about all the livestock producer stories shared in *Resilient Agriculture* and where you can find them.

Vegetables

A.G. Kawamura
Orange County Produce, Fullerton, California

I would say that our weather is both "predictably unpredictable" or "predictably predictable." We're generally dry from April all the way until December. We'll get a few storms every now and then, a kind of monsoon that comes up the coast in the fall, but generally we have some of the most predictable weather anywhere on the planet. In my experience, nothing's changed that much.

— A.G. Kawamura

When A.G. Kawamura heads out each morning to check on his crops, his route is an unusual one for a vegetable grower. He drives into residential developments, onto military bases and through city parks, schools and abandoned orange groves to get to his fields. His family didn't set out to be urban farmers, but they started farming early enough and stayed in business long enough that the city eventually grew out to reach them. "We are definitely urban producers or farmers in an urban area," A.G. says. "It was a rural area when we started farming here. The city came to us and then it surrounded us. We've just never left."

A.G.'s grandparents came to southern California from Japan around the turn of the last century and made their living in the agriculture of their new home. They did whatever work they could find in those early days, one set of grandparents picking and packing oranges, sharecropping and landscaping, and another grandparent starting a small fertilizer and farm supply company. After the Kawamura families were released from an Arizona internment camp in 1945, they returned home to the Los Angeles area to rebuild their lives. Over a decade later, the family moved farm operations to Orange County, growing and shipping produce in the area, which was well-known at the time for growing oranges, walnuts, tomatoes, lima beans, and asparagus, along with other vegetable and horticulture crops.

Credit: A.G. Kawamura

A.G. Kawamura

As the area population grew, rising costs and skyrocketing real estate prices forced many Orange County growers to sell out. Those that remained continued to grow on ground leased from several large private landowners and military bases. "We don't own any of the ground we farm on here in the county," A.G. explains, "and that's a challenge because we rent ground from the utilities, from a school district, from cities and counties, from the military and from private developers. We will farm any vacant lot that's over four or five acres. If I can see the weeds are growing well, and I can see that there's a fire hydrant or recycled water connection nearby, then we look at those as viable places to farm."

Today, A.G. and his brother Matthew are the third generation at Orange County Produce (OC Produce) to grow and market fresh fruits and vegetables along and within the urban boundaries of Orange County. A.G. manages production and coordinates production schedules with the company's network of growers and Matthew manages distribution, sales and finances. A.G.'s nephew, Paul Kawamura, became the fourth generation to enter the produce business when he launched GEM-Pack Berries, a new company that expands OC Produce's strawberry production and sales to a year-round venture. The expansion required new marketing and financing which was strategically provided by their longtime growing partners. Together, A.G., Matthew and Paul work to continue their family's legacy by producing and distributing regionally grown fresh fruits and vegetables produced with a commitment to healthy land, people and community.

Using climate smart and certified organic practices, A.G. has grown strawberries, green beans, cabbage, summer and winter squash, celery and other produce crops on about 1,000 acres of leased land within and near Orange County for more than 40 years. Some practices that he views as key to successful growing include improved crop cultivars, cover crops and compost, integrated pest management, water conditioning, drip irrigation and precision agriculture techniques to prepare ground and plant, water and feed his crops.

"Our fresh market growing season here starts basically in January and goes to about May," A.G. explains. "After our season is done, we work with partners to the north all the way up to Watsonville. Then into the fall and winter, we head down into Mexico to work with partners in Michoacan." These partnerships with other growers provide OC Produce with a consistent, year-round supply of produce sustained by relationships cultivated over many years. "We've been working with our green bean grower in Mexico for over 25 years," says A.G., "and one of our strawberry growers for more than 60. We're a classic grower-shipper. We're not a co-op. We agree to handle other growers' products along with our own under our various labels. It's all based on a handshake."

OC Produce's urban production model keeps A.G. constantly juggling land that is more suited or less suited for organic farming. The difficulty obtaining long-term leases on land limits his ability to expand organic production, which typically accounts for about 30 percent of his production. "We have pieces that we think will be there for the rest of my productive life, even though we don't own them, and then we have pieces that I might lose after the next crop. We have a lot of very short-term leases that limit us from making certain investments in soil that we would otherwise do when we know we have longevity. So, our farming runs the gamut, but I would say we're definitely sustainable…as long as we have access to land." A.G. uses recommended crop production practices on all of his acres and describes his management philosophy as "feed the soil to feed the plant kind of thinking."

According to A.G., soil health, water quality and quantity, pests and disease, labor and regulations are his most challenging long-term management concerns. These challenges are shared by produce growers everywhere, but increase in complexity in urban farming, especially when continually shifting to new ground from year to year. "We are always struggling with very weak to almost dead pieces of ground. Several times we've walked onto land that can take us two years of work before we can actually get anything to grow. Depending on the structure of the lease, we work with the landowner to bring those properties back into good production. We've also been lucky to find some pieces that were just incredible. Tremendous fertility, tremendous biological activity in the soil."

It was farming one of these parcels many years ago that convinced A.G. of the sustainability benefits of healthy soils. "Our focus on soil goes back to that period of time when we were farming on pretty tired soils using basically an all-chemical regime. We thought the yields we were getting were okay. Then we got a chance to lease an abandoned pasture at a Future Farmers of America high school. The soil had been pasture land for 20-something years. It was deep, rich, black soil with earthworms and the whole thing. We planted a crop of green beans and it was just amazing. That really opened my eyes to the value of feeding the soil."

Like most growers in California, A.G. is always thinking about water— access, quantity, quality and cost—but continuing drought in California has pushed the demand for water past the breaking point on some parcels. "Over the last decade, I had to abandon one parcel that was close to 100 acres as the well went dry. Over several years, we kept cutting back on planted acres and then finally just pulled out completely. We had no choice because there was no plan B, we had no secondary source of water." A.G. makes use of several strategies to secure the water he needs for crop production. He has invested in new water-conditioning technology to reduce the harmful effects of rising salt concentrations in the region's surface and groundwater supplies. He also looks for opportunities to lease new parcels in areas with access to both high-quality reclaimed water and potable water.

While critical to the success of his operations, these water management strategies add significantly to A.G.'s costs of production. "We've stopped growing some crops because of the cost of water," A.G. explains. "The further south you go in California, the more expensive the water becomes. It is not unusual for water to cost $2,000/acre for a crop of strawberries on the low end, and double that if your only supply is expensive potable water. The rising costs for rent and water throw us out of whack on a bunch of products that we used to grow, and that we can grow very well in our climate. Celery is the best example. I don't grow celery anymore. Not because we can't grow it, but because its cost structure is too high, especially in a conventional market. It's just too high."

Managing field labor is another common challenge in produce production and one that has grown increasingly complex as agricultural labor regulations have changed. "The more highly perishable the crop, the more intense your issue is with labor," says A.G. "You might be late in harvesting some other crops, and so what, but in the crops that we grow, every day makes a difference, hours can make a difference. We've been fortunate that we haven't had extreme labor challenges." The shift from seasonal to year-round production at Orange County Produce has reduced the risk of labor disruptions because A.G. can offer field workers full-time employment on his farms, other farms in the supply network and in the packing and distribution side of the business. Many employees have been with the company for decades.

Farming in the southern California climate brings its own set of challenges that have not changed over time, according to A.G. "I would say that our weather is both 'predictably unpredictable' or 'predictably predictable.' We're generally dry from April all the way until December.

We'll get a few storms every now and then, a kind of monsoon that comes up the coast in the fall, but generally we have some of the most predictable weather anywhere on the planet. In my experience, nothing's changed that much." Frequent periods of high temperatures and record heat waves threaten crops in the region, but A.G. successfully manages these risks by selecting locations based on weather records and trialing drought- and heat-resistant crop cultivars. "Improved weather forecasting models and soil moisture monitoring opens the door for much greater irrigation efficiencies during stressful climate events," says A.G.

The region's infamous Santa Ana winds are one of A.G.'s most difficult production challenges. "They are created by a high pressure zone that gets set inland and pushes tremendously high winds here, 50, 60, 70 mile an hour winds sometimes," A.G. explains. "We usually get tougher winds than most of the rest of the country. They're not quite hurricane force, but they cause havoc, and they cause problems for whatever you have growing. When they come in, all we can do is keep our crops irrigated and then basically hope it stops."

More recent adjustments to weather-related risks at OC Produce have been driven by new state regulations, rather than regional changes in weather. "Of course we have always taken precautions," A.G. says. "We have rest periods and we provide shade and water for our field crews, but California created new heat stress protections for farm labor that require us to shut down operations at temperatures that we would have typically worked in before." A.G. anticipates having to make some additional adjustments in production when the state's new water management restrictions come into effect.

Thinking back on his family's business success over the years, A.G. values the resilience benefits cultivated by OC Produce's success in marketing. "We were not really good growers when we started out," says A.G., "but we have always been good at quality of product and marketing. We became better at growing because we had to in order to succeed. We've continued to be pretty good at marketing and that's why we're still in business. It doesn't matter how good you are at growing, if you don't have good marketing skills you will soon be out of business. Some of the best growers that I know are not farming anymore and that's really sad."

Many years of growing in a growers' network also yields resilience benefits. "There's nothing that can replace experience and observational knowledge of a region," says A.G. "Actually knowing your area and being able to recognize vulnerabilities or opportunities. We share and exchange information across our companies and we work collaboratively to help each other. As a good example, we had a disease complex I couldn't identify. I texted a picture of it to our partner down in Mexico and he immediately ID'd it as a new mildew and gave us a whole new prescription on how to handle it. It was a perfect example of how quickly I can find some help when I'm dealing with something I haven't seen before. We must keep expanding our knowledge base."

Meeting southern California's need for a consistent supply of high-quality produce using the company's urban production model has honed A.G.'s ability to successfully navigate a complex and constantly shifting working landscape. It has also taught him how to swiftly restore degraded soils in urban and suburban areas using farm-scale practices. This kind of resilience thinking will become increasingly valuable as he explores new opportunities to sustain food and farming in these times.

"Over the 40 years that I've been farming," says A.G., "we've moved our equipment yard at least six times, which is kind of unheard of for most farmers, but it's just the way we farm. We're moving around a lot. I don't have a single place that I'm farming on today that I farmed on 18 years ago. So we've focused on investing in technology and equipment that improves the predictability of our outcome. Drip irrigation, injection systems, filter systems, satellite precision planting and leveling, mechanical harvest, cover crops, innovative compost programs and improved varieties. These are the kinds of growing tools that help us get the unpredictability out of our system."

Thinking about the future, A.G. is confident that these kinds of tools will help to sustain agriculture no matter the challenges ahead. "The key to our continued success," A.G. says, "will be to embrace change and to evolve with the help of innovative collaborations that allow us to produce food in unconventional places and create working landscapes that deliver multiple benefits for the community and for the farmer."

A.G. Kawamura has been a leader in local, state and national agricultural organizations for many years. He serves as the co-chair of Solutions From the Land, a nonprofit organization that promotes collaborative, farmer-led climate smart land management strategies in support of the UN Sustainable Development Goals, is a founding member of the Agricultural Sustainability Institute at UC Davis, is on the board of Western Growers and served as the Secretary of the California Department of Food and Agriculture from 2003 to 2010. A.G.'s service to California agriculture was recognized by CalPoly Pomona with the 2017 Jim Hicks Agricultural Achievement Award and his leadership in support of the U.S. food and agriculture industry through education and rural development was recognized with a Leader in Agriculture award by Agriculture Future of America in 2017.

Hannah Breckbill
Humble Hands Harvest, Decorah, Iowa

I actually thrive in change and that's really good for adapting to the unexpected. Emily is a very diligent and very forward-thinking planner. She thinks of worst-case scenarios and she plans for them. So those two personalities together are able to deal with whatever is coming in different ways. When one of us is struggling, the other one usually has it covered.

— Hannah Breckbill

Humble Hands Harvest is a worker-owned cooperative farm growing organic vegetables, organic grass-fed and finished lamb and pastured pork on 22 acres in northeast Iowa near Decorah. Hannah Breckbill is founder and co-owner of the farm with her second cousin Emily Fagan. Over the last decade, Hannah has worked to cultivate the resources, skills and experience needed to own and manage a successful farm business through participation in the Land Stewardship Project's Farm Beginnings and Journeyperson programs and the Practical Farmers of Iowa's Savings Incentives Program.[1] Along the way, she innovated a new cooperative model of farm ownership that cultivates the resilience of community-based food and farming.

Hannah Breckbill and Emily Fagan

After graduating from college with a degree in mathematics, Hannah was drawn to farming because it offered an opportunity to combine her passion for social activism with her love of the land. After working for other vegetable farmers in her first three growing seasons, Hannah established Humble Hands Harvest in 2013 and continued to farm on leased land in several locations in the Driftless area of southwest Wisconsin and northeast Iowa. Hannah celebrates the experiences of each new season and location—both good and bad—as important steps towards her goal of establishing a permanent farming enterprise. "I've grown a lot through running my own farm business," says Hannah, "but in order to really begin caring for the land, really investing in soil building and perennial crops, I needed a permanent place. The question was, how will I attain that?"

In 2014, Hannah participated in a cooperative purchase of a farm near Decorah that was initially motivated to protect the land from development.[2] As a shareholder in the farm, Hannah successfully encouraged the owners to shift their goal from farmland protection to farmland access. In doing so, she found the answer to her question. Using her own savings, a family loan and matching funds earned through the LSP's Journeyman's Program, Hannah purchased eight acres of the farm in 2016. A year later, Emily joined Humble Hands Harvest as a co-owner. Since then, Hannah and Emily have worked together to raise the funds needed to develop vegetable and livestock operations on the farm through events like a farm-raising party and a Go Fund Me campaign.

Today, Hannah and Emily produce diversified vegetables, lamb and pork using mostly hand labor on about two acres of cultivated ground and 20 acres of pasture.[3] Vegetable production infrastructure on the farm includes a well, a drip irrigation system, a deer fence, a moveable high tunnel, a greenhouse, a cooler and a pole building that includes a packhouse and storage. They own a small tractor they use to cultivate ground for cropping and to mow pastures and headlands, manage a small flock of sheep to produce grass-finished lamb and finish about ten feeder pigs each year on pasture. Hannah and Emily direct market their vegetables and meat cuts at twice-weekly farmers markets in Decorah and through on-farm sales of lamb and pork halves and wholes.

Thinking back over years Hannah has been farming in the Driftless area, she's found soil and water management to have been especially challenging regardless of location. Years of conventional corn and soybean production in the region have eroded topsoil, leaving heavy clay subsoils on the surface. Besides being infertile and physically difficult to manage, water management on such degraded soils is especially difficult.

"The land that we're on now was conventionally cropped corn and beans and very erodible land," says Hannah. "There was basically no topsoil when we got onto this place, so it can get really waterlogged, and then you can't really work the soil at all." Having the resources to secure enough committed, experienced labor has been a third significant management challenge.

In her relatively short time farming, Hannah has experienced a number of major weather-related challenges. During the extraordinary 2012 drought, the farm she was working on went under, and she was let go in the middle of the growing season. Four years later, her crops were destroyed by record flooding in the Iowa River watershed. "I was renting land in 2016 next to the upper Iowa River. The river rose 14 feet in late August after an 11-inch overnight rain and covered my bottom field. I lost absolutely everything I was growing that year. That was rough."

Hannah believes weather patterns have changed over the time she has been farming, most notably in the fall and spring. In her early years of farming in the Driftless region, she became confident that she could reliably establish winter cover crops in the dry fall weather. In 2016, the pattern seemed to change. "It just kept pouring, just inches at a time all through September and into October," Hannah recalled. "It was just absurd. And ever since then we've had wet falls."

Hannah has also noticed that "soggy and cool" spring weather seems to be more common making it harder to stay on schedule with spring planting. "Even just five years ago, it seems like there would just be a lot more days in March in the 60s. I think people around here would agree with me that it's been more overcast in the past few years than it used to be." Although summer weather seems about the same, Hannah does think that over the time she has been farming, heavy summer downpours have become more common.

These changes in weather have made it more challenging to get field work done on time. "I have a policy that I've stuck with very well through my farming years," Hannah says. "I don't do field work or farm by headlamp. I just don't do it. And we don't work on Sundays." Wet weather all spring in 2019 put Hannah and Emily so far behind schedule that a short dry period forecast for a Sunday into Monday required a temporary change in policy. "We knew we needed to plant," recalls Hannah, "so we just busted it for 12 hours that Sunday. I was putting butternut squash in by headlamp at 10 p.m. that evening. So yeah, there is a feeling these days that you've got to make hay while the sun shines, as they say."

Hannah and Emily have reduced the risks associated with more wet weather in the spring and fall by adding some protected growing space at Humble Hands Harvest. "I hadn't been a person to grow in high tunnels or that kind of thing," says Hannah, "but we do now have a couple caterpillar tunnels and a high tunnel." The physical protection has made it much easier to manage disease during rainy periods in summer and the extra warmth under cover seems to improve the growth of summer crops like peppers, tomatoes and melons.

Difficulties staying on schedule with spring planting also got Hannah thinking about adapting their current field management practices which rely on tillage to prepare ground for planting crops. At a Practical Farmers of Iowa (PFI) winter vegetable growers meetup in 2019, Hannah was part of a discussion about no-till as a possible solution to the challenges of managing heavy clay soils in wet weather. Soon after, Hannah learned about Singing Frogs Farm,[4] a small-scale, no-till vegetable operation in California. She was inspired to give it a try. "At the beginning of April, I just took a shovel to the field and made four 150-foot raised beds with pathways in between. It turned out to be a really wet spring and those four beds were the only places that we could plant for a really long time."

That fall, Hannah and Emily made ten more raised no-till beds as a way to learn more about how best to manage the practice on their farm. They planted the beds with an oat cover crop that winter-killed and applied compost the following spring right before planting. After some more tweaking and a third spring of testing in 2021, Hannah and Emily have decided to expand their area in no-till. "We like it and we are thinking it is the way forward," Hannah explains. "It's strange, though, because both this spring and last spring have been quite dry, meaning that our original reason for using no-till is moot. But we've started using silage tarps a few weeks in advance of planting on the no-till beds to reduce weed pressure and it's been great! And plants—cucurbits in particular—really react well to the heavy compost application."

Taking a step back to think about the farm resources that contribute the most to the climate resilience of her business, Hannah points to soil health, crop diversity, a complementary management partnership, cost-share opportunities and the knowledge and skills she has gained through her participation in community-based research and education programs.

Hannah appreciates the complementary contributions of the co-farmers' personalities to the farm's solid management team. "I actually thrive in change," says Hannah, "and that's really good for adapting to the unexpected, like new weather norms or COVID. Emily is a very diligent and very forward-thinking planner. She thinks of worst-case scenarios and she plans for them and is really good about that. So those two personalities together are able to deal with whatever is coming in different ways. When one of us is struggling, the other one usually has it covered."

Business planning at Humble Hands Harvest is grounded in the holistic management practices that Hannah and Emily both learned in farmer training programs. "Farm Beginnings starts with Holistic Management," Hannah explains, "and I really appreciate that way of looking at things. It has been very valuable to me. We have a mission and vision statement that we go back to and we have communication processes that we feel are really important."

Humble Hands Harvest has also benefited from community-based technical assistance offered by organizations such as PFI and cooperative extension and they have taken advantage of USDA agricultural conservation technical and financial assistance programs managed by local Natural Resource Conservation Service (NRCS) staff to help fund their farm startup. "NRCS cost share helped us buy our high tunnel and renovate our pasture," says Hannah, "and we have some local foods extension people who are fabulous." Hannah and Emily have also received funding to do on-farm research from PFI and North Central Sustainable Agriculture, Research and Education program. Although they have explored the possibility of government-subsidized insurance, it doesn't seem to be a good option for their operation.

Hannah also views their choice of production scale and marketing as a resilience benefit. "Since we're direct market farmers and we do CSA," Hannah explains, "the timing of crops isn't super-urgent. So as the season seems to be kind of moving, we can just move with it. I feel like we're at the right scale and we're doing the right kind of marketing to be able to deal with weather changes."

Hannah points to the ability to pivot overnight in response to the COVID pandemic as another example of the resilience benefits of their choice of marketing and scale. As communities began shutting down and grocery store shelves were emptied in the first few weeks of the pandemic, Hannah worked with other local growers to pilot a new cooperative marketing and distribution program that aggregated products from different farms for home delivery. "It's been really exciting and it feels good," says Hannah. "It feels like we're being recognized by more people in our community as a valuable resource. One of the things I'm thinking about with COVID is how do we take what we've learned and take it forward, even when we get back to so called 'normal'? What can we learn about the local or regional food system that will serve us well going forward?" The new marketing program returned in 2021, now managed by the Iowa Food Hub and expanded to include communities outside of Decorah.

Working cooperatively with Emily and supported by her community, Hannah feels confident that she has the intellectual and emotional resources, along with the farming knowledge and skills, needed to keep Humble Hands Harvest on the path to success despite the inevitable challenges ahead. Although she has achieved the permanence that she wanted, she remains passionate about helping other beginning farmers achieve their own dreams.

"The issue that feels really important to me right now is land access and ownership," says Hannah. "I look at the landscape in Iowa and think about how much of it is in corn and beans produced by tenant farmers on land owned by people who live far away. This kind of system is not good socially for farmers or rural communities, and I don't think it's good ecologically for the planet under climate change either. I want people to understand that it is possible to start a farm from scratch and how that possibility relies on using unconventional systems."

Hannah Breckbill is a passionate advocate for rethinking sustainable land tenure for young and beginning farmers, organizes for local resilience to the climate crisis, is the founder of the Queer Farmer Convergence and eagerly seeks out opportunities to encourage new ways of thinking about food and farming in community. She regularly leads community-based actions, workshops and other events on her farm and throughout the Midwest. Hannah was recognized by the Midwest Organic and Sustainable Education Service (MOSES) for her work as a farmer-activist with a Changemaker Award in 2021.

Rebecca Graff and Tom Ruggieri
Fair Share Farm, Kearney, Missouri

This year may have been an average year in the end, but what really happened was that we had a big rain event and then no rain for a month and then another big rain event, so we got 12 inches in July and then nothing in August and September. So, if you just look at the average, it looks like an average year, but we basically seem to be fluctuating from drought to flood and back again.

— Tom Ruggieri

Rebecca Graff and Tom Ruggieri own and operate Fair Share Farm, a diversified vegetable farm located in rural Clay County, Missouri. Together they manage about ten acres of annual and perennial vegetables and fruits, culinary herbs and a large flock of laying hens to feed people in the Kansas City metro area. They market fresh vegetables, fruits and eggs from their farm through a CSA and produce vegetable ferments for direct wholesale and farmers market sales. They also coordinate with other producers in their area to offer meat, cheese and bread options to their CSA members. This farm-based food hub/food circle model allows them to provide a more diverse group of products to their CSA while also cultivating a diverse network of local food production capacity.

Rebecca and Tom met back in 2001, just a year after each had left unsatisfying jobs to look for a better way to support healthy community by working in sustainable agriculture. Their paths crossed at Peacework Farm,[5] an organic vegetable farm in western New York where Rebecca was working as a first-year apprentice. "Tom came out to one of the first member workdays," Rebecca recalls. "It was mid-May and I had been there just about a month. I admired his leek-trimming skills as we were preparing vegetables for market. We went on our first date a week later. We apprenticed together the following year and we've been farming together ever since."

Credit: Fair Share Farm

Tom Ruggieri and Rebecca Graff

After another year working as apprentices at the Micheala Farm in southeast Indiana, Rebecca and Tom headed back to Missouri to begin farming on land that has been in Rebecca's family for four generations. The Graff family farm, like most farms in the area, produced corn, soybeans and cattle using conventional commodity production practices. "I grew up on a farm," says Rebecca, "but I did not learn how to grow crops of any kind, so it was all new to me when I started my first apprenticeship." Using holistic farm planning[6] and biological farming practices, Rebecca and Tom worked over the years to create a farm sustained by healthy soils and healthy community.

They credit their use of holistic whole farm planning with keeping their business on course towards the vision that brought them together at Peacework. "I think we're where we are because we are both going in the same direction," says Tom. "Rebecca sees a little farther down the road than I do sometimes and that helps a lot. She has a little more vision and I think you need that. It can be hard for two people to work towards the same vision, let alone really spend the time to say what's really important to them. We both feel pretty strongly that what we're doing reflects our principles. We haven't changed them over the time we've been farming and that's exciting, especially this day and age."

A key holistic planning practice is a written three-part holistic goal that articulates the farm family's quality of life goals, different options for how the farm can support a desired quality of life and a vision of how the farm and the community that it serves must be far into the future in order to sustain that quality of life. "We revisit our holistic goal any time we're making decisions to change how we're doing things," Rebecca explains. "It definitely came into play with COVID when we had to work fast to change how we interact with the public."

Some key soil health practices that support Tom and Rebecca's vision for Fair Share include crop rotation, reducing tillage, cover cropping, compost application and extensive mulching, the production of vegetable ferments and rotational grazing with their flock of a hundred laying hens. The soil health benefits of most of these practices are widely recognized, but the production of vegetable ferments is less common. "We believe the fermented vegetable operation complements our efforts to improve the life and health of our soil," Tom explains. "It also diversifies our operation. We encourage more growers to explore this low-energy demand, live-culture option for adding value to their vegetables."

Tom and Rebecca began to notice soil behavior changes within the first five years of implementing these practices on what was formerly conventional corn and soybean ground. Since 2008, soil organic matter content in their cultivated fields has increased by more than two percent. Among the many benefits of the soil health practices used on the farm, this increase in soil organic matter suggests that on average, the farm sequesters over 50,000 pounds of carbon per year.

More generally, Tom and Rebecca have put a focus on looking for ways to enhance the ability of their business as a whole to conserve and restore natural resources. They reject the use of synthetic insecticides, herbicides, fungicides, fertilizers, GMOs and chemically treated seeds and manage their land to enhance biodiversity and create beneficial habitat for wildlife. They have also put a focus on reducing greenhouse gas emissions with a sustainable infrastructure plan that includes participation in a Green Power program and have recently installed a grid-tied

20kW solar array. Their greenhouse is a passive solar design, is partially earth-bermed and uses 50 black, water-filled barrels for passive heat storage. Other infrastructure improvements include a solar-powered irrigation system and an electric utility vehicle. They also converted their Allis-Chalmers G tractor from gas to electric and power it with a 48-volt battery pack. Rebecca and Tom estimate that these improvements have reduced the farm's carbon footprint by more than 400,000 pounds since 2004.

Whole farm planning has helped the couple overcome several long-term management challenges at Fair Share. The farm is situated on a sloping, highly eroded ridge with thin topsoil and a heavy clay subsoil—a combination of landscape characteristics that has made water management particularly challenging. "Water has been our biggest problem," says Rebecca. "We've lost a lot of crops to stagnant water conditions. Not having a well-drained soil and trying to grow annual vegetables is a bad combination. So we've done a lot with compost and cover cropping to try and bring up our organic matter levels and that has helped." Other long-term challenges include managing pests and diseases and securing reliable, experienced labor.

Over their time at Fair Share, Rebecca and Tom believe that seasonal weather patterns have changed, although they are quick to point out that it is hard to tell because of their location in a region well-known for weather extremes. "We're on the edge of the Great Plains here," says Rebecca. "This part of the country has always had a certain amount of wild fluctuations because we are influenced by weather coming up from the Gulf, across from the Southwest and down from the North. There's so many different things happening that it's easy for our weather to change dramatically."

Even so, they agree that extremes of temperature and moisture seem to be more common now than when they first started farming. "A couple of times this past fall," says Rebecca, "we went from 85 degrees to a freeze in 24 hours." These extremes can be hidden in weather averages. "This year may have been an average year in the end," explains Tom, "but what really happened was that we had a big rain event and then no rain for a month and then another big rain event, so we got 12 inches in July and then nothing in August and September. So, if you just look at the average, it looks like an average year, but we basically seem to be fluctuating from drought to flood and back again."

Over the last decade, Rebecca and Tom have made a number of adjustments to reduce the risks from these weather changes, particularly more variable spring weather and higher spring and summer temperatures. They have shifted their tomato planting to later in the spring to reduce the risk of frost/freeze damage, switched to heat-resistant leafy green crops and tomato cultivars and use more mulch to protect cultivated fields from heavy rains during the growing season. "We've changed our tomato planting schedule to later," said Tom, "because there's no reason to rush them out there. And we've had springs where it's been 90 degrees for two weeks on end in late April and May and spring crops like napa cabbage and lettuce just literally melt in the field." Rebecca adds that shifting production out of the hottest summer months is the most common adaptation being talked about among the vegetable growers that they know.

Another change to make it easier to manage field crops in wet weather has been the purchase of a new field cultivator that reduces the need for tillage. "We till a lot less than we used to," says Tom, "and when we do, it is mainly to kill and incorporate cover crops. We bought a new field cultivator with a leveling bar on it and a rolling harrow. We mow down the cover crops, cultivate lightly to incorporate and let it all compost together. After that, we do as much as we can to not turn the soil over." Rotational grazing with the laying hen flock also helps with cover crop management. "We didn't really get the chickens for weather-related conditions," says Rebecca, "but they've really helped a lot with managing cover crops, particularly when it is wet. We can put them in an area and let them scratch everything down and then you don't have to till, you can just run the cultivator through there and that gives us a lot more flexibility to get in and out of a field."

Rebecca and Tom have also shifted to completing field preparation for planting in the fall and covering prepared ground with tarps until needed for planting. This system allows them to stay on schedule with spring planting no matter the weather. "We've started to use tarps more, silage tarps, and that has been beneficial for us," says Rebecca. "We set up our fields in the fall and then just pull the tarp off and plant in the spring. We also do this for the summer and early fall plantings because it seems to protect the soil from losing moisture in the heat of the summer."

Increases in heavy rainfall in the last five years have forced Tom and Rebecca to rethink their water management strategies at Fair Share. Heavy rainfalls have damaged or destroyed crops and have required them to take some areas out of production because of increased surface runoff and soil erosion. They have resisted installing tile drainage—the most common strategy used by farmers in their area to move excess water from crop fields—and have looked instead for a more sustainable alternative.

In December 2019 they attended a workshop on Restoration Agriculture by Mark Shepard[7] where they learned about Master Line design for water management. Master Line design uses earthworks to effectively slow, spread and soak water throughout a landscape to increase soil organic matter and improve farm resilience to both flooding rains and drought. The Master Line system is an adaptation of P.A. Yeoman's Keyline Design for use in temperate environments.

In the fall of 2020, Rebecca and Tom installed over 7,000 linear feet of berms and swales across their field production areas, pitched downhill at a one-percent slope and spaced 40 feet apart. This layout created alleys where they can farm annual or perennial crops or run livestock. In 2021, they began the process of planting the berms with perennial trees and shrubs, including chestnut, persimmon, pawpaw, elderberry and hazelnut. They plan to add a flock of sheep in future to help with land management and further diversify the farm.

Tom and Rebecca point to four key assets that support climate risk management at Fair Share Farm: irrigation, soil health, a long partnership based on careful observation of ecological processes on the farm and a willingness to work together to find solutions that support their values and goals. "Definitely irrigation is one of our biggest assets," says Rebecca, "being able to keep crops watered through droughts. We have a pond on the farm that my family built. That's been really critical for us because we usually have at least a month, if not more, where we may not get any rain in the summer."

Tom appreciates the way that their quest for a sustainable water management solution changed their thinking about their land. "Doing the Master Line design helped us realize that we had not been looking at the farm as three dimensional. When we first laid out our fields, we laid everything out in rectangles, as if our ground was flat. But it isn't flat, it's contoured hilly ground. It's as if we had bound up our land in ill-fitting clothes in our original layout, so it could not fully respond to challenges like a heavy rain. The swales and berms just fit the land better."

Although the Master Line design will take some years to fully develop, a heavy six-inch rain just after they had finished installing the berms and swales provided immediate benefit. "We still had runoff and ponding in some areas," said Tom, "but there was an interesting change in one area which produced our best crop of napa cabbage ever." "It's early days," Rebecca said, "we'll have to work with it awhile to see how it develops."

Tom and Rebecca are unsure about the success of their farm, and of farming more generally, as climate change creates new challenges for growers everywhere. They agree that regenerative vegetable farmers need a better network of support to increase the climate resilience of their operations. "I think what has to happen first," Tom offers, "is that reality has to take root in our nation. We have to accept the fact that climate change is real." Rebecca adds, "I think that a lot could be done if there was more of a push nationally, in our state and locally towards sustainable agriculture, but it seems like that is something the people in power don't want to promote or encourage."

Rebecca Graff and **Tom Ruggieri** have been actively involved in community-based sustainable agriculture research and education projects and programs with their CSA members, farm apprentices and other farmers for many years. They serve as a mentor farm in the Growing Growers Kansas City program and regularly teach workshops on biological vegetable production, CSA marketing and vegetable fermentation at agricultural conferences in the Midwest.

Marc White, Keymah Durden and David Hester
Rid-All Green Partnership, Cleveland, Ohio

> A year ago, we had this polar vortex, when it was minus 40 degrees for a week and a half straight. Then, this year we're in late November and it's 60 degrees. Early this year, it rained almost all spring and all early summer. It's so unpredictable. So we have to be very adaptable to extreme weather changes and, excuse me, it's sad to say, but we know it's not going to get any better, it's going to continue to get worse.
>
> — Keymah Durden

From the Public Square in downtown Cleveland, the trip out to the farm is an easy ten minutes by car on a Saturday morning. Head southeast on Ontario St. and continue onto Orange Avenue, then take a slight left onto Woodland Ave. and head due east out of town. Take a right on Kinsman Road, a left onto 81st St., then a right on Otter Ave. and you've arrived at the Rid-All Green Partnership's city farm. Although just a short drive from downtown Cleveland, decades of disinvestment have left the Kinsman area so isolated that it was nicknamed The Forgotten Triangle. It's a problem faced by many post-industrial cities in the Midwest. Factories closed and when the white middle class took flight into the suburbs, they took investment capital, new industries and jobs with them.

Credit: Gary Yasaki

David Hester,
Keynah Durden,
Damian Forshe,
Randy McShepard
and Marc White

The Rid-All farm began as a vision shared by three men who grew up together nearby, left home to find their fortunes in other places and returned to give back to the community that raised them. They came home to transform the land and people that they loved while sharing a gospel of soil: heal the soil, heal the people, heal the community, heal the planet.

"Our thing has been to see how far we can push this," says Marc White, a founding partner and project manager. "To see how much we can do as urban farmers to help our community. It's profound to see the effects of what we have created grow and change on a daily basis, year after year, season after season." Marc's many years in fashion design inform his work as the farm's general manager and lead value-added product developer. He uses his design experience to create beautiful landscapes on the farm that produce healthy foods designed to promote the health and beauty of the people that enjoy his Urban Farm Doctor's line of regenerative superfoods juice and food products.

But the landscape wasn't always beautiful or the soils healthy. Initial testing showed that the soils on-site were burdened with the toxic legacy of the area's slow decline from a thriving African-American neighborhood in the 1950s to a burned-out wasteland that eventually became a community dumping ground. The first step on a path to healing was the removal of tons of trash and contaminated soil.

"In the first phase of site reclamation, we learned the soil had high levels of lead, arsenic and other heavy metals," said Marc. "So that was excavated and we brought in some Ashtabula field soil, then laid down a barrier, then layered four to five feet of wood chips over the whole area." The excavation and then fill of clean soil plus wood chips served to jump-start the development of a healthy new ecosystem on the site. "This is a huge aspect of what we teach needs to be done in the city," Marc explains. "Repair the soil first and simultaneously as we work."

With the help of a growing group of volunteers and funders attracted to the founding partners' vision, the site remediation was followed in a few short years by aquaculture greenhouses, community gardens, educational programs, hoop houses, a farmers market and a CSA. Today, the Rid-All Green Partnership supports more than 20 diverse social enterprises that produce and market compost, bedding plants, fresh vegetables and fish, medicinal herbs and beverages, as well as market materials like wood chips, field dirt and gravel that others need to establish their own community gardens.

The partners agree that the farm's most important product is educated young people who are reconnected with their own farming heritage, skilled in urban growing and excited to share the gospel of soil. Drop by the farm just about any day and you will likely find 15 or 20 gardeners, volunteers, students and youth education leaders working alongside staff in the farm's many enterprises. Promoting multigenerational learning to cultivate the sharing of different skills and experiences between elders, youth and ancestors is central to Rid-All's educational philosophy.

"Our commitment since day one has been to get our youth excited about putting their hands in the soil," says Keymah Durden, co-founder, environmental educator, engineer and educational program manager at the farm. "What our youth bring to the table is not just a willingness to learn what we know about growing or the traditional knowledge that we learned from our grandparents. Our youth bring with them a whole wave of new things, like social media savvy, that us older folks don't touch as much. Our way of teaching closes a loop, because we are able to reach everyone in the community."

The Rid-All farm covers about 15 acres and includes an industrial composting facility, a community garden, a small orchard with a diverse mix of fruit tree species and cultivars and about 40,000 feet of covered growing space in four hoop houses and two aquaponics greenhouses. Much of the fresh produce marketed by the farm is grown undercover in the hoop houses and the entire site is landscaped with a diversity of perennial vegetables, fruits, nuts and culinary and medicinal herbs.

"When you enter our site," says Keymah, "you enter into an oasis. We strive to captivate you with the beauty of plants and then we invite you inside our various structures to show you what's possible, in terms of the production of fruits, vegetables, fish and herbs." The Rid-All farm also markets fresh, nutrient-dense field-grown greens and other vegetables that are staples in the African-American community in collaboration with Amish farmers growing in Ashtabula County—a rural area near the outskirts of Cleveland about 30 minutes from the farm.

Rid-All direct markets most of its produce through an on-site farmers market and a unique CSA program. "We partnered with Groupon and Amazon Prime to sell one- to six-month CSA memberships," Keymah explains, "which gives members access to our vegetables, our fresh produce, our farm-raised tilapia fish, soil products and training services. This continues to work out very well for us."

In 2020, the Partnership jumped on an opportunity to develop a retail store in the nearby neighborhood of Maple Heights. They are particularly excited to bring a fresh produce market back to a community that has not had a supermarket selling fresh, nutrient-dense and culturally-appropriate foods for decades. "What started out as just a local food stand at our farm," says Keymah, "has grown into a seven-day, 24/7 operation where we're selling healthy food, produce from our farm and farm partners as well as produce from other local distributors."

With the exception of vegetables produced outside in summer, most of the produce grown on the farm is grown undercover. "We only have three months of a summer season to really grow stuff outside," says founding partner Dave Hester, a.k.a. "Dr. Greenhand," an experienced horticulturist who manages crop production, "but we grow year around in the hoop houses." Even in such controlled environments, Dave, like farmers everywhere, points to weeds as a major challenge on the farm. Long-term weather-related challenges typical of the region include extreme cold in winter, late spring frosts and the disease challenges like powdery mildew that come along with hot and humid summers. Labor, always a challenge in farming, is made more difficult because the farm depends primarily on the labor of students and volunteers working with a small group of staff and youth leaders.

Even though much of production is under cover, the partners have noticed changes in weather patterns that have offered both challenge and opportunity. More variable temperatures and dramatic temperature swings, unseasonable temperatures, extreme winds and weather have created some challenges since the farm was established in 2010.

Dave has noticed that more variable spring weather, particularly high temperatures early in the season, seems to confuse many perennials growing on-site. "The changing weather patterns seem to give a false signal to the plants, especially if the nice weather lasts more than three days." He goes on the explain that an unusual "warm snap" in the early spring starts plants growing and they keep growing even after temperatures drop again. "I guess they have a problem figuring out what season it really is," says Dave. He has also noticed what might be some weather-related disturbances in tree fruit pollination on the farm.

In an effort to address pollination issues, Dave has made of point of including more flowering plants on site for the farm's bees and other pollinators and he has seen some improvement in recent years. "The bees and butterflies seemed to have really thrived this past year," said Dave. "They are staying on the property more which is always a good thing, because they just do so much. But I would say that generally, we just don't know what the weather's going to be like at any given moment." Increased weather variability has also interfered with the composting enterprise, which requires regular periods of dry and relatively warm weather to support the composting process.

Extreme weather poses unique risks to the farm's indoor growing space. "We are a 24/7 operation," says Keymah, "in that every day somebody has to check on the farm, whether it's to see if the power's off, see if the wind has damaged something, check that the heaters or the ventilation systems are operating properly." Power outages are a major concern on the farm, because even short periods without power at critical periods in the development of plants and fish can significantly reduce yields or even cause death.

As weather variability and extremes have grown more intense, the partners have made a significant investment in backup generators. "They are not cheap," says Keymah. "They range anywhere from $50,000 to $75,000 for a generator system that can run our aquaponic tanks. When catastrophic weather hits, we have to be able to respond very quickly because two hours without any power could mean we lose 10,000 fish that took two years to raise. That's a real bad day and we've experienced it."

Keymah goes on to explain that weather extremes seem to be more common. "A year ago, we had this polar vortex, when it was minus 40 degrees for a week and a half straight. Then this year (2020),we're in late November, and it's 60 degrees. Early this year, it rained almost all spring and all early summer. It's so unpredictable. So we have to be very adaptable to extreme weather changes and, excuse me, it's sad to say, but we know it's not going to get any better, it's going to continue to get worse."

The partners have been incorporating climate-smart and regenerative agriculture practices into their operations and their educational programs over the past few years. They have also begun to explore the use of biochar production to heat their greenhouses in winter and produce a new marketable product. Other adaptations that draw on both old and new practices include using high-insulation plastics to cover indoor growing spaces, upgrading irrigation systems, heating in winter with compost and meeting some power needs in the aquaponics operation with on-site solar.

"We have developed a system inside our hoop houses to keep them warm in winter with what we call a 'hot mix,'" Dave explains. "It's a combination of wood chips, beer hops, coffee or tea grounds, plus some worm castings sometimes, which will heat the hoop house up by 20 to as much as 40 degrees, depending on outside temperatures and the size of the hoop house. So we're able to keep it warm enough to where you need to take your coat off when you go inside, even when there's a foot of snow outside."

Stacking practices like the hot mix and high-insulation covers has worked well at the farm. "We grow a variety of tropical plants in our main greenhouse," says Dave, "including plants like orange and avocado. They do fine even in our lowest winter temperatures."

While these practices are important to the climate resilience of the farm, the partners agree that it is the commitment of the people involved in the farm that is key. "The most important thing we have going for us is a really good team that has gotten to know each other and to grow with each other," says Dave.

Keymah agrees. "You can have all of the fancy technology you could ever pay for, but your greatest asset is always going to be the human resource, because the situations that occur are hands-on. You've got to get out of bed and get in a car and go down to the farm and see what happened. You can't Google it. You can't Instagram it. You can't Facebook it. And you can call these guys 24/7 and they'll go, they'll respond." Marc adds, "There's no digital agriculture," to smiles and nods of agreement all around.

As innovators and educators, the partners know they need to be forward thinking about how climate change changes what they do and how they do it. "We know the fact is that climate change has a negative impact on growing food," says Keymah, "but if we are flexible enough, we think we can turn that negative into a positive. Most of the experts talking about climate change have pointed to solutions that are Earth-based. They're not abstract. They're not scientific or highly technical. They're the things that our ancestors did hundreds of years ago. They're Earth-based solutions that we can employ, not just to secure the food system, but to secure the future of life on the planet."

A good example of this kind of forward thinking is the farm's newest enterprise. As a key partner in Cleveland's new program to reforest the city as a climate resilience strategy, the partners are busy making plans to grow and distribute 300,000 tree seedlings to city residents over the next five years.

Reflecting on how the farm has changed since they moved that first load of contaminated soil, the partners agree that they are thankful to be working on the leading edge of food and farming solutions. "We figured that what was missing within our community was a sense of agriculture being foundational and fundamental," said Marc. "We knew from the start that what we are doing with Rid-All was not just growing food. We are growing relationships, we are growing community, we are growing love. We are growing so many things besides just some food, you know? This is the gospel of the soil, and it's a whole 'nother level. Because out of the soil comes all this opportunity."

The **Rid-All Green Partnership** was co-founded by Keymah Durden, Randy McShepard and Damien Forshe in 2011 with the help of founding partners Marc White, Dave Hester and Timothy Lewis and the guidance of Will Allen and Growing Power. The Partnership's achievements over the years have been made possible by the Cleveland city officials, foundations and other nonprofit groups and many individuals who have supported the Partnership by sharing good ideas, materials, labor, funding and friendship over the years.

Ira Wallace and Mary Berry
Southern Exposure Seed Exchange, Mineral, Virginia

> We've had more crop failures in the last few years due to heavy rains and flooding than we've had in all of the last 20. It's the biggest thing we hear from our growers and it's the biggest pain here at Acorn as well. Many of our growers have had complete crop failures from flooding in two out of the last three years.
>
> — Ira Wallace

Acorn Community is an egalitarian, income-sharing, farm-based community located in south central Virginia near the town of Mineral. The community's 20 members collectively own 72 acres of farmland, woodlands and wetlands including 50 certified organic acres that produce food for the community and vegetable seeds for the Southern Exposure Seed Exchange, the community's nationally-recognized cooperative seed business.

All Acorn community members work part-time on the farm and in the seed business. Ira Wallace has coordinated seed production and processing at Southern Exposure for about 27 years. Mary Berry joined the community in 2018. She works with Southern Exposure's customers and co-manages seed and food production for the community. She has about three years of growing experience.

Southern Exposure direct markets more than 800 varieties of vegetable, flower, herb, grain and cover crop seeds produced at Acorn and by a network of more than 70 small and mid-scale farmers located mostly in the Southeast. The company's co-owners also produce educational materials and support a diversity of programs to further the company's mission to democratize the seed supply, promote organic agriculture and gardens, and preserve the genetic heritage and diversity of southern food and farming. "Our marketing strategy is more about education," Ira explains. "Over the years, rather than have a lot of ads, we write stories, and blogs, and white papers, and work with other organizations to educate people, with the aim that, as people learn more about organics and heirloom plants, they will try our seeds and become customers."

The community values farming practices that reduce the need for fossil fuels and other purchased inputs and promote healthy ecosystem services to produce healthy crops. Seed growers at Acorn use certified organic practices on about four acres each year, including about a quarter acre of protected growing space in three greenhouses. Key practices at Acorn include long rotations that include both summer and winter cover crops, the application of compost and the use of organic mulches. Cover crop species that have worked well at Acorn include buckwheat, rye, clover, vetch, southern peas and sunn hemp.

Irrigation, weeds, pests and diseases, and timely field operations are some key long-term management challenges at Acorn. The farm's irrigation difficulties are mostly related to inadequate infrastructure. The community has just one well to supply water for both domestic and farm use. "During a long, dry period," Mary explains, "it can be problematic to get that water everywhere that we need it to go, especially when we

Credit: Acorn Community

Ira Wallace and Mary Berry

only have a limited amount of it." The community has explored adding a second well on the property, but the best well sites are located far from the community's center and the most heavily cultivated fields.

Managing insects and disease in seed crops is another longstanding challenge. "Seed plants are in the ground a lot longer," Ira explains, "so you are likely to encounter pests like cabbage seed pod weevils that you never see when you are producing vegetables. It can be hard to find good information about management of vegetable seed pests." Soil-borne diseases are a particular concern. "As seeds people, we must think a little bit further along the line than farmers and gardeners," says Ira. "We want to be sure to keep our products disease free." Practices that reduce the risk of pest and disease damage at Acorn include long crop rotations, a focus on soil health and shifting the production of seed crops threatened by disease at Acorn to other farms in Southern Exposure's production network.

Balancing labor needs between production, processing and sales at Southern Exposure has also been a challenge over the years. "Weed management is one of the things we aren't doing as well as we used to," Ira explains. "We've been successful building up the seed company which means we have more need for office-type work. We don't hire outside of our community, so we have to figure out how best to use the available labor." One strategy that is working well for Acorn is shared-labor partnerships with other nearby intentional communities to meet time-sensitive, high-labor seed production needs such as weed management.

More generally, the dual role of the farm within the greater Acorn community adds complexity to managing labor on the farm. Planning crop successions and production volumes to meet both seasonal community food needs and business goals requires some sophisticated seasonal and annual crop planning. "When you live in a place like Acorn," Mary explains, "there's so much going on all the time. It can be hard to just figure out where to put your labor. Sometimes anything that is not a seed crop, like food crops for the community, are neglected when things get really busy in the business."

Thinking about changing weather patterns over the years, Ira remembers that the weather was fairly predictable during her first decade farming at Acorn. She started to notice an increase in extreme rainfall and more frequent dry periods and drought around 2005. "If you look at the average over the years," Ira explains, "it's not such a big change from year to year. But averages don't tell you what is actually happening. It's crazy. It won't rain for six weeks and then in the next three weeks you get enough rain to make it the average for that period." Mary finds it interesting to hear Ira say that weather patterns used to be more predictable at Acorn. "For me, weather has always been this big chaotic thing," says Mary. "At the beginning of each year, I'm like, 'All right, what kind of year do I think it's going to be?'" Ira confirms Mary's perception, noting that "as long as you've been farming, I think it's true, it has been this way."

Other weather patterns creating new production challenges at Acorn include increased temperature variability and extremes, higher temperatures, heat waves and a lengthening but more variable growing season. Ira feels certain that temperatures are warmer, particularly in the fall. "We used to tell people they could start planting garlic and perennial onions in the middle of September," she explains, "but the sun is usually too hot at that time now. We've shifted our planting recommendations to the middle of October and November." As Southern Exposure's customer service rep, Mary sometimes works with customers who are having trouble making the shift to later fall planting. "People call me in September wondering when we will send their garlic order. I look up weather patterns in their area with them to help them understand it's still too early. Often times their response is still, 'Well, I always plant garlic in September.' And then I tell them, 'The times are changing.'"

A longer growing season presents both opportunity and risk at Acorn. "Generally speaking," says Ira, "it seems like we should be able to take advantage of the longer growing season, but weather variability in both spring and fall also seems to be growing. Some years, it's warm much earlier and falls do seem to be getting longer, but in any given year, you can't tell what's going to happen." Summer and winter weather also seem to be growing more unpredictable. "We are seeing long drought periods," says Ira, "longer than average on record. And cold weather, always a fleeting thing where we are, is hard on plants when 30-degree temperature drops happen repeatedly."

Acorn's growers have adopted new practices like no-till to reduce the risks of more heavy rainfalls, dry periods and drought. "Occultation,[8] in particular, is big and new for us," Ira says. "We work with a seed grower in Austin, Texas, who was a Texas water steward of the year. They do a small amount of hand-watering when they transplant, and otherwise, they don't irrigate. They manage their soils to capture all the rain that falls and then slowly release it to plants. It made me think that if they could do that in Austin, we could do it here." The purchase of a new no-till transplanter is Acorn's most recent step into no-till. "We do a lot of transplanting because of our need to keep track of all these varieties," Ira explained. "The folks at Living Energy Farm[9] told us about a human-powered no-till transplanter developed in South Africa. So we imported one."

More variable weather has required increased attention to the timing of harvesting and processing seed crops at Acorn. "Knowing when the seed is close enough to maturity to harvest, even when they haven't fully dried up, is more important now," Ira says. "We've learned we can't leave seed crops to dry down in the field like we used to, because we never know when we might get a heavy rain. So we harvest seed as soon as we can and do the final drying indoors."

Weather-related disruptions in Southern Exposure's growers' network seem to be increasing as well. "We've had more crop failures in the last few years due to heavy rains and flooding than we've had in all of the last 20," says Ira. "It's the biggest thing we hear from our growers and it's the biggest pain here at Acorn as well. Many of our growers have had complete crop failures from flooding in two out of the last three years." Growers throughout the network are making adjustments to reduce the risk of increased weather variability and many are actively working together to identify and develop flood-, heat- and disease-tolerant crop cultivars.

Both Ira and Mary recognize some new opportunities created by changing weather patterns, especially in the lengthening growing season and warming temperatures. Mary notes that unlike more experienced growers, her thinking about what might be possible is not limited by past experience. "One year I planted watermelons very late in the season," says Mary, "and more experienced gardeners told me 'Oh, it's too late, you're not going to get anything.' But I ended up with a great crop of watermelons. It's not that I think they don't know what they're talking about, but I do think that outside factors have changed since they've acquired their knowledge. I see it as a sign that maybe people need to start trying out new things." Ira agrees, adding, "Even though I have the records to prove that the growing season is longer, I don't always act like that's true."

The longer growing season and warming temperatures have enhanced Ira's success working with tropical perennials. "A lot of our tropical perennials have been brought by people from all over the world by different ethnic groups," says Ira. "I'm particularly interested in things from the African diaspora. This work is not just about promoting traditional plants, it is also about retelling the story of agriculture, starting with African people being brought here because of their experience as farmers and not just as brute labor in the Southeast." Southern Exposure is currently leading a national project to recover and share traditional collard varieties.[10]

Mary and Ira both value the resilience cultivated by life in an intentional community that welcomes a constant flow of visitors who bring lots of enthusiasm for trying out new ideas. They also appreciate working within a large network of Southern Exposure's seed growers, plus farming and gardening customers, who are eager to share their experiences of growing in a changing climate. "Since I've started growing, I've learned that you can't just rely on the things you know now," Mary explains. "You've got to be willing to adapt and change and learn, constantly. It's the way of the future."

If weather changes grow more intense, both Ira and Mary expect that Acorn's growers will most likely continue to emphasize soil health practices, add more infrastructure such as irrigation, drainage and protected growing and seed processing space, and put more emphasis on trialing robust cultivars for drought and heat tolerance. Ira imagines that costs of production will likely increase on the farm and in the seed business, while Mary imagines greater efforts to rethink traditional crops and innovate new growing practices. Both are very interested to explore landscape-scale options to improve water management. "We have some new farming neighbors," Ira says. "They're permaculture people. They did a lot of landscape work, like putting in swales, before they started to farm. It's amazing how quickly they got into production and now after ten years, their place is amazing."

Thinking about the future, Ira has confidence that Acorn and Southern Exposure will continue to be successful as a community and as a business. "Here's the thing," Ira says. "The things that we can do as individuals and as farmers to make our communities more resilient, such as being more welcoming to pollinators and increasing biodiversity and so forth, are all things that also help with climate change. So for me, working with growers is our little piece of making it possible for all of us to eat 20 years from now." Responding to Ira's optimism, Mary says, "I want to have hope, but I don't know. There's also the reality of how we're treating the planet right now. I think we are all going to be a little lost in the dark as the weather keep getting more and more chaotic."

Mary is excited by the potential of changing not only how and what we grow, but what we choose to eat. She finds hope in research that promotes eliminating meat from our diet as a powerful climate change solution with a multitude of other environmental and social benefits. "If we want to stop climate change, making the switch to a vegan lifestyle is an option that is open to a lot of people," Mary says. "I know this is an uncomfortable topic for a lot of people, but I think we all need to be willing to go outside of our comfort zone a bit more to try and save the planet."

Ira agrees with Mary's point, but invites us all to think just a bit more broadly about the need for fundamental change in the food system. "Having spent years being both a vegan and a vegetarian at different times," Ira says, "I don't think that really gets to all of it. For me, the number one thing that needs talking about in the U.S. is that 'cheap food' is not cheap. Food has to cost more than what we pay for it now in order for human beings to work in the food system with dignity. This is one change that we don't talk about enough."

Mary Berry joined the Acorn community in 2018. **Ira Wallace** is a founding member of the community. She is a regular contributor of essays and articles to *Mother Earth News, Fine Gardening*, and Southern Exposure and teaches workshops in Southeast culinary history, the production of heirloom and open-pollinated seeds, and organic gardening and farming methods in the Southeast and beyond. She serves on the boards of the Organic Seed Alliance, Organic Seed Growers and Trade Association, and the Virginia Association of Biological Farmers. Ira was nationally recognized in 2019 as a Great American Gardener by the American Horticultural Association. Her book, *The Timber Press Guide to Vegetable Gardening in the Southeast* was published in 2013.

Pam Dawling
Twin Oaks Community, Louisa, Virginia

We've tried leaving more things to overwinter because it doesn't get as cold as it used to, but it's a bit trial and error because we just never quite know what's going to happen. Winter weather in Virginia is all over the place. It just seems more that way recently.

—Pam Dawling

For more than half of her nearly 50 years of farming, Pam Dawling has grown food at Twin Oaks, an intentional community and ecovillage of about a hundred people located in central Virginia near Louisa. Pam managed vegetables and fruit production on the community's organic farm, which also produces dairy, beef, poultry, honey, herbs, tree fruit, mushrooms, seeds, ornamental flowers and forestry products. The garden produces a diverse mix of vegetables and berries on about 3.5 acres of cultivated fields, raised beds and undercover in a high tunnel. Crop rotation, cover crops, and the application of compost, plus careful attention to season extension have been key production practices in the market garden. In 2017, Pam's role shifted from manager to support staff at the garden, but she brings her full range of experience at Twin Oaks to this story.

Pam Dawling

Pam's early experiences as a member of World Wide Opportunities on Organic Farms introduced her to organic farming as a healthy way to live—as well as to the difficulties of solo farming—and sparked her interest in living and working collectively. Growing for a community food supply allowed Pam to center vegetable production on crops with high food value, rather than high market value. She also managed production to include specific crops favored by community members and to meet community needs for vegetables throughout the year. To achieve these goals, Pam focused production on a select group of crops: leafy greens like lettuce, chard and kale that can be grown year-round using some combination of field and protected growing space; summer crops that are equally tasty fresh or processed such as tomatoes, squash, cucumbers and peppers; and root crops like sweet potatoes that are easy to store with minimal processing for use throughout the year.

Thinking back over the last three decades, Pam identifies labor, weed pressure and timely field work as the most challenging aspects of growing vegetables at Twin Oaks. While these challenges are common to farming everywhere, the use of community members for labor adds additional management complexity. "We get a ton of people with little experience," Pam explains. "This means that sometimes the work is not done to the highest standard." Years of dependence on community labor have led to high weed seed populations in the soil and contributed to longstanding difficulties with getting field work done on time. "Some community members are very enthusiastic and will be there reliably," says Pam. "Others are more fair-weather gardeners who are less likely to understand that something needs to be finished by a certain time. It takes a bit of experience to appreciate the bigger picture and to be motivated by the need to get this done today because the rest of the week we've got to do this, that and the other."

Long-term weather-related challenges at Twin Oaks include heavy rains, drought and summer heat and humidity. The garden's clayey subsoil is slow to drain and standing water after a heavy rain is common. Although the garden had insufficient irrigation during long dry periods or drought for many years, recent investments in a new well and drip irrigation have helped to improve water management. The high heat and humidity typical of summer in central Virginia create difficult working conditions. "Some of our workers just don't want to work outside when it's hot and humid," says Pam, "and for those that do, it's a challenge to drink enough water and just keep going."

Pam has noticed a number of changes in weather patterns over the time she has grown vegetables at Twin Oaks. Heavy rains, dry periods and droughts, and extreme temperature swings seem to be more common. Winters are warmer and summer temperatures seem to be increasing. All of these changes have made managing longstanding weather risks even more challenging.

"We get more high temperature days in the summer," Pam says. "I'm talking over 95 degrees for a high. That is tough. It's tough on the people and it's tough on some of the crops as well. It's hard on the people because you've got the workload and you need to get out there and do it." Limiting field work to the cool morning hours is complicated by disease management considerations. "The dew can be quite heavy in the mornings, so we try and leave those crops alone until the leaves dry off. But then you really don't want to be out in the field in the heat of the afternoon. It makes getting the harvesting done difficult."

It seems to Pam that precipitation patterns are also changing. "I'm not sure if we've had more, longer drought periods than we used to," says Pam, "but I think maybe we have. We used to have more regular rains. We used to have late afternoon thunderstorms pretty regularly and we haven't really had those for the past few years." She appreciates the improved water management made possible by the new well and drip irrigation. "We are better set up for irrigation now than we used to be," says Pam. "So the dryness hasn't been so much of an issue. Now we have drip irrigation for some crops and the overhead sprinklers for when we direct sow a crop."

Warming winters and more variable winter temperatures are increasingly interfering with crop management at Twin Oaks. "We've noticed with the cover crops that we're probably going to not be able to rely on oats to winter kill too much longer," Pam says. Although temperature swings in winter can be expected in central Virginia, "it seems like the temperatures zigzag up and down more than they used to," Pam explains. "We don't have the kind of winters when you think right, it's winter, everything needs to be wrapped up and done. Winter weather in Virginia is all over the place. It just seems more that way recently." Pam also sees some opportunity in warming winter temperatures. "We've tried leaving more things to overwinter than we used because it doesn't get quite so cold," says Pam, "but it's a bit trial and error because we just never quite know what's going to happen."

More variable weather over the last decade has required Pam to put more focus on being prepared to complete field work swiftly when conditions permit. "I found that we really need to snap to it when conditions are right to get things done," Pam explains. "It seems like it has become more risky to say 'Oh, yeah, maybe I'll do that tomorrow instead.' It's best, if the conditions are right, to just get the job done—just in case." Pam has also come to appreciate the flexibility of protected growing space to reduce the risks associated with more variable weather.

Pam thinks that the lessons learned during her long experience growing at Twin Oaks are an important contribution to the resilience of the garden. "We have been working on learning to forecast the weather ourselves," says Pam, "I mean, to be able to look at the sky and figure out what's likely to happen. I think having more of a weather sense is going to be more necessary. I've also been doing more recording soil temperatures because I think it will make a big difference in knowing when to plant things, rather than relying on calendars to guide field work."

Other variables that Pam monitors daily or weekly in the market garden include temperature, precipitation, key phenological benchmarks and pest and disease populations. "When I do my once a week walk around," Pam explains, "I am sure to go everywhere and look at everything under every row cover and in every corner to be sure there are no surprises about to burst on the scene. I think this is very worthwhile, because it is so easy to get focused on urgent tasks and lose sight of the important ones. To be a successful farmer, you've got to find a way to balance the importance and the urgency."

Another critical skill in these times is the ability to manage field work in more variable conditions. "We have a monthly calendar with a task list. We take a look around once a week and say, 'Right, let's make a list for the week.' But you can't just start at the top of the list and work down. Each task involves some judgement about matching field work to current crop and soil conditions. For example, say it has just rained or it's going to rain, so it's a good time to do some transplanting, or we are going into a hot spell, so we need to put shade over a crop. Besides just doing the work, you have to factor in what has happened and how to respond along with how to prepare for what's likely to happen. That's a lot to keep in mind all at once, it's a big juggling game like keeping all the plates spinning. It's a skill that takes some learning."

More generally, Pam thinks about farm resilience as a question of increasing management flexibility. "Mostly I am thinking about how to be more nimble and more attentive to details on the farm," Pam explains. "I'm thinking about how to keep our options open, about diversifying crops to spread risk, about keeping the soil covered and about having backup plans." For example, she has started managing transplants a bit differently in the last few years. "When we start transplants," says Pam, "we keep some in reserve in case something goes wrong. It used to be that we just kept some extras if we had them, but now we will bump up extras into bigger pots and keep them thriving for a week or two in case we need to do some replanting."

If these changing weather patterns continue to grow more intense in coming years, Pam imagines additional changes will be needed to sustain vegetable production at Twin Oaks. More protected growing space will be important and the community may need to consider adding drainage or using more raised beds to reduce the production risks associated with heavy rains. Pam also wonders about how occultation might be used to improve soil conditions. "I'm just reading Daniel May's book about no-till farming," Pam explains, "and thinking that using tarps might really work well in the garden, both for the weeds and for protecting the soil from heavy rains. Although you've got a bit of a problem, if you have a lot of runoff from a plastic tarp. There's maybe more to think about on that one. I do wonder if the way of the future here at Twin Oaks may be less tilling and more manual work with small scale equipment."

Pam appreciates the role that a supportive community of farmers has played in her personal resilience as a farmer. "I have found a lot of value in having local groups of farmers and growers to share experiences and to brainstorm ideas with," Pam says. "It is so valuable to have peers and mentors to take your worries to who can give you some fresh thinking. This includes less technical, but more emotional support, a shoulder to cry on. Equally important is to have some domestic support.

Someone who cooks your dinner if you have to stay out late finishing out something."

Pam is excited about the potential for new farmers to find this kind of support in the growing number of farming education programs. "I've been appreciating secondhand the beginning farmer programs that are around these days that didn't use to be," says Pam. "I think that's a very valuable resource for new farmers. Offering education and mentorship in solidarity with other growers is very valuable both for learning growing skills and also for mental health. This is such a hard time to be a new farmer."

Pam Dawling regularly leads vegetable production workshops at sustainable agriculture and organic farming conferences and Mother Earth News Fairs, writes technical and general interest articles for gardening and farming organizations, produces a weekly Sustainable Market Farming blog, is a contributing editor with *Growing for Market*, and is the author of two best-selling books, *Sustainable Market Farming* and *The Year-Round Hoophouse*. Since her retirement in 2017, Pam has continued to support the Twin Oaks organic farm as a farm advisor, manager of vegetable transplant production and as a year-round hoop house laborer. Stepping away from daily farm management has given Pam more time to teach, consult, speak and write about sustainable gardening and farming.

C. Bernard Obie
Abanitu Organics, Roxboro, North Carolina

I can say that for us, the main ingredient is being connected to the Earth and the forces, angelic and otherwise, that govern the growth of plants. We will accept what nature gives us and we'll be grateful. We'll do our part to try to keep it in balance and healthy. I think in the long term, that's really all we can do as Mother Nature will have the first and the last word about our existence here.

— C. Bernard Obie

C. Bernard Obie, known as "Obie" to his family and friends, grows certified organic vegetables, fruits and herbs on his family farm in north central North Carolina near Roxboro. Obie founded Abanitu Organics on the land that has nourished his family since his great grandmother, Ms. Lucy Obie, purchased the farm in 1906. Lucy's son, John, was a widely respected farmer in the area. Obie's father, Bernard, continued the family's farming legacy, even as all of his siblings left the farm for city jobs. "His was a really pivotal decision," says Obie, referring to his father's decision to become a farmer, "because he could have very well done the same thing. For a lot of folks, that's the time when the connection to the land was broken."

C. Bernard Obie

Obie's earliest memories include helping out with farm chores along with his six brothers and sisters. "I call Person County 'the land that time forgot for a while,'" says Obie, "because we were still doing things in the 1950s the way folks had done them a generation earlier. We did not have a tractor, we still ploughed with mules. Our cash crop was flue-cured tobacco. We kept hogs and chickens, had a milk cow and grew corn and wheat for our own food and to feed our animals. That's how we came up." Most of the farm's produce was for family use, but the tobacco and hogs were sold for cash income. Obie values the unique perspective on life gained by growing up on his family's farmstead.

Obie's parents placed a high value on their children's education and encouraged all of their children to go to college. "Everybody graduated from college," Obie recalls, "which meant that everybody left the farm." Obie earned a degree in animal science and went to work in the agricultural pharmaceutical industry after graduation. "It was a backstage pass into meat production," says Obie, "and all the horror associated with that. The wholesale slaughter of animals for food was something that I could not find a place for in the way that I understood life and health and wellbeing." Obie left pharmaceuticals for a job in the health insurance industry, an experience that raised his awareness of the ill-health of many Americans. "I was determined to find out why people were so sick," Obie said, "and the conclusion I came to was that food was the main culprit."

Around the time that Obie was connecting the dots between industrial agriculture and human health problems, a family tragedy called Obie back to the farm to manage the aftermath of his father's unexpected death. "I came home to deal with all the stuff that has to happen when the head of the family dies suddenly," Obie recalls. "We had to sell off all the farm equipment to settle debts, but we were able to keep the land." Obie stayed on to help his mother with the farm. He founded Abanitu Organics in 2006 and left his job to farm full-time in 2008.

"So that's how I got back into farming," says Obie. "I wanted to help otherwise ordinary folks connect the dots like I did to understand why they were obese, why they had hypertension, why they were diabetic and so forth." Obie was particularly interested to work with people, especially poor people, who did not have access to good food. "I wanted to get good food into the hands of folks, so that they could taste it and feel it and then draw their own conclusions about why good food matters."

Obie is the fourth generation to care for his family's 92 acres of mostly old mixed hardwood forest, some meadows and gardens that are home to Abanitu Organics. He cultivates about 12 certified organic acres on the farm, including five to six acres of diversified vegetables, one acre of asparagus, a 1.5-acre orchard and an acre of muscadine grapes. Obie uses a mix of organic and biodynamic crop production practices including crop rotation, cover crops, insectary plantings and the application of purchased compost and biodynamic preparations. Field work on the farm is guided by the Stella Natura calendar, a management tool used by biodynamic farmers to plan fieldwork. Produce is marketed primarily through weekly farmers markets in Roxboro and Durham and CSAs serving members in Roxboro, Durham, Charlotte, Columbia, South Carolina, and Northern Virginia.

Although Obie puts an emphasis on creating healthy growing conditions on the farm, he knows that conditions are not always ideal. "Our soil is heavy red clay and full of feldspar. We're working with what we have and doing the best that we can to take the work forward. So we ask the plants to do their best in our circumstances."

According to Obie, relating in this way with the plants in his care is every bit as important as any particular farming practice that he uses. "One of the things that I understand is that plants are sentient beings," Obie explains. "They sacrifice themselves for our well-being and they're conscious and aware. So we try to engage them as such. We sing to the plants, we talk to them, we explain to them why they're being called into the world, what their mission is. We ask them to help me to understand them better, to be better partners in doing this work of healing." In return, Obie promises to help his customers respect and appreciate the contributions that plants make to their health and their lives.

Abanitu Organics grows a variety of fresh market vegetables year-round, but producing a diverse mix of cool-season nutrient-dense leafy greens is a major focus of the farm. "Coming from a place of wanting to impact most profoundly people's health," Obie explains, "we believe that nothing does that as much as leafy greens. Nothing has that range and that power and diversity. Nothing offers so many healing properties: anti-oxidant, anti-inflammatory, anti-cancer, promotes elimination and detoxification, supports heart health and brain function, and strengthens the microbiome. So for us it's greens and it's been greens from the beginning."

Long-term production challenges at Abanitu Organics include wildlife damage, irrigation management and access to experienced labor. "Deer and groundhogs in particular are probably the biggest challenge that we've faced," says Obie. "We've responded by putting up fences, a solution that is expensive and time consuming and requires a fair amount of work but it has worked for the most part." Delivering water when and where it is needed can be a logistical challenge at times. "We've got some infrastructure still to put in place to make it easier for us to irrigate in extreme conditions," Obie explains. "If we hadn't had to do so much with fencing, we would have upgraded our irrigation quite a while ago."

Like many vegetable growers, meeting changing labor needs over the years has presented some challenges at Abanitu. "We've had different experiences, different levels of attitude and productivity from some of the folks that we've hired," Obie explains. "We know we need to do something different and we're looking now at what that could be. Whatever it is, we want to treat people fairly but at the same time get what we need in terms of production from folks. That's what we're struggling with and that's the next big thing."

Heat waves, damaging storms and heavy rains have been consistent challenges over the last 15 years or so. Obie manages the production risks associated with high temperatures and heat waves with a mix of practices including adaptive planting, physical protection and careful attention to irrigation. Severe storms create risks to critical infrastructure on the farm. "If a tree falls on fencing," Obie explains, "then that's work that has to be done again. That has happened. Let's just say the severity of some of the storms is an issue, probably for everybody."

The farm's heavy clay soils make the production risks created by heavy rainfalls particularly difficult to manage. "A heavy rain presents an issue with getting into the field to work," says Obie. "The soils just don't dry out quickly and there is a tendency for plants to be stunted or even killed in low-lying areas because moisture sticks around so long in our soils." Obie built a set of raised beds to improve his ability to adapt to wet soil conditions. "In a year where there's a lot of rain and it's early and the temps are cold, the raised beds give us some additional options," says Obie. He also notes that the challenge of farming in clay is a good example of the "give and take" of farming, because although clay soils are slow to dry out, they also provide some welcome drought tolerance.

Asked about changing weather patterns, Obie thinks back to when he was a child working with his father on the farm. "I can remember periods of time when weather threatened the crop," Obie recalls. "It does seem like the frequency and maybe the severity has picked up a little bit since then, but it's always been an issue here to one degree or another."

Winters also seem to be getting warmer. "The old folks say they used to get a lot more snow," says Obie, "and I can remember that as a child. It seemed like we got more consistent cold weather. I have vivid memories of the extreme cold, the snow that came up over my knees. We don't get much snow now. That's different. I don't know why though." Obie has also noticed a lengthening growing season that is "warmer longer and I'd even say warmer sooner too."

With Abanitu Organic's focus on producing cool season leafy greens, Obie has had to make some adjustments to these changes in weather. "Just timing the field operations, I'd say that is the biggest thing for us. Greens really don't like the heat that much. The earlier spring makes it more of a question mark. Especially things that depend on the cool to develop and grow fully and powerfully. So in the spring, we try to get out as quick as we can to avoid a back-end heat wave." Along with the earlier planting, Obie has also started to stagger planting dates as well. "When we started putting things out earlier, we also started to hedge our bets by putting some plants out, then we'll wait for the next good sign day to put some more stuff out." Obie's comment about a "good sign day" refers to his use of the Stella Natura calendar which uses biodynamic principles to identify the best days in each month to work with crops that are harvested for their roots, fruits, leaves or flowers.

Warmer weather in the fall has also required some adjustments at Abanitu. Although hot August temperatures make it difficult to establish fall plantings, waiting for cooler weather to plant is less and less of an option as the growing season lengthens. For now, Obie has been able to establish fall plantings by watering regularly and shading the young plants with row covers. He also uses the Stella Natura calendar to guide his planting schedule when conditions are difficult. "When you plant under a good sign," Obie explains, "the plants have a better chance of doing well. If everything goes smoothly and conditions are perfect, then it doesn't matter that much, but if there's adversity, it seems to me that plants appreciate that extra energy."

Obie has also begun to exploring the potential of warm season leafy greens like Swiss chard, Malabar spinach and callaloo,[11] a green native to the Caribbean. "We planted callaloo for the first time last year," says Obie, "and I'm over the moon about its potential for our farm. Loves the heat, loves the humidity and it grows. It has so much power. It rivals or exceeds the conventional cruciferous greens in what it brings to the table."

Obie views his relationships with his crops as the most important resilience asset on his farm. "We try to prepare plants for what it is that we're calling them into the world for," says Obie. "We engage them to help us with our mission and we promise them we will do our best as well." Although Obie recognizes that some biodynamic practices like the Stella Natura calendar are "not scientific," his experience confirms that working according to the calendar is helpful. A second practice that Obie uses to cultivate resilience on the farm is "just taking the time when you're out in the field to observe, to be quiet and just be in the space to get a feel for how things are going."

Other vegetable growers in Obie's area have adapted to these changing weather patterns by moving to physical protection for all of their crops, either in high tunnels or with plastic mulches. "I have a little resistance to that," says Obie, "but that's just me personally. I guess it's working for them, but I can't get past all the plastic. At the end of the day, our intention is to grow food, but we want to grow it in a way that honors the Earth. In a way that honors the plant. In a way that says we're partners in creation."

Thinking about the future, Obie is confident that he has the resources to manage the risks associated with changing weather patterns. "I can say that for us, the main ingredient is being connected to the Earth and the forces, angelic and otherwise, that govern the growth of plants," says Obie. "We will accept what nature gives us and we'll be grateful. We'll do our part to try to keep it in balance and healthy. I think in the long term, that's really all we can do as Mother Nature will have the first and the last word about our existence here."

C. Bernard Obie has been active as a leader and supporter of healthy food and farming projects and programs in North Carolina over the last two decades. Through Lucy's Phratry Farm, LLC, he brought three generations of his family together to heal themselves and protect their land. He is a member of the Rural Advancement Fund's Farmers of Color Network, the Ausar Auset Society International and the Southeastern African American Farmers' Organic Network. Obie regularly leads community-based workshops in healthy food and gardening practices, has served as President of the Durham Farmers Market, and has written about health, food and farming for *Ebony Magazine*.

Table 15.1. Resilient Agriculture Vegetable Producer Stories

Review this list to learn more about all the stories shared by vegetable growers that are featured in the 1st and 2nd editions of *Resilient Agriculture*. The stories marked with an asterisk are included in this volume. Updated stories from the 1st edition, and links to more information about all of the *Resilient Agriculture* producers can be found at realworldresilience.com

Farm or Ranch (Region)	Scale	P	A	T	Featured Resilience Behaviors
Nash's Organic Produce (NW)	1000	X	X		Add equipment, double scale, drop fresh produce/on-farm grocery.
Full Belly Farm (SW)	400	X	X		Improve irrigation, farming seasonal edges, riparian restoration.
Orange County Produce (SW)*	1000	X	X		Urban soils restoration, precision management, growers' network.
Rockey Farms (SW)	500			X	Shift some cash to cover crops, insectory intercrops, livestock.
Monroe Organic Farms (SW)	105	X	X		Improve irrigation, add protected space.
Fair Share Farm (MW)*	10	X	X		Carbon farming, vegetable ferments, Master Line earthworks.

Table 15.1. (cont'd.) Resilient Agriculture Vegetable Producer Stories

Farm or Ranch (Region)	Scale	P	A	T	Featured Resilience Behaviors
Humble Hands Harvest (MW)*	22	X	X		Cooperative land access, worker-owned farm, no-till raised beds.
Harmony Valley Farm (MW)	200	X	X		Social recovery reserves, add cooling, retreat from floodplains.
Rid-All Green Partnership (MW)*	15	X	X	X	Urban soil/community restoration, 20 social enterprises, improve recovery reserves, growers' network, carbon farming.
Peacework Farm (NE)	20	X	X		Add wells/irrigation, shift work to cooler hours, agricultural justice.
New Morning Farm (NE)	45	X	X		Add protected growing/organic pesticides, floodplain retreat.
Southern Exposure Seed Exchange (SE)*	50	X	X		Add protected space, no-till raised beds, growers' network.
Twin Oaks Community (SE)*	4	X	X		Add protected growing space, monitoring.
Abanitu Organics (SE)*	12	X	X		Add no-till raised beds/heat-tolerant crops, shift growing season.
Perry-winkle Farm (SE)	10	X	X		Intensive cover cropping.
Peregrine Farm (SE)	5	X	X		Improve water capture, shift growing season, shift to heat-tolerant cultivars, drop sensitive species.
Maple Spring Gardens (SE)	14	X	X		Shift growing season, drop sensitive species, shift to heat-tolerant cultivars, add protected space.

Note: Scale is reported as acres under management; P, A, T refer to types of adaptive strategies used by the producer: Protect, Adapt and Transform (see Fig. 11.1), respectively.

16

Fruits and Nuts

Mark Shepard
New Forest Farm, Viroqua, Wisconsin

Since we've been here our longest drought was two calendar years where we had snow in winter time, but almost zero measurable rain in the summer. Then in 2018 and 2019, we had twice the annual amounts of rainfall. Seventy-five inches of rain one year to zero inches of rain the next year. That's a challenge.

— Mark Shepard

When Mark Shepard and his family first visited the land that would become New Forest Farm almost 30 years ago, they looked out over a Midwest landscape of degraded croplands typical of late twentieth century industrial agriculture. Gazing across the treeless property covered in empty corn and hayfields, the Shepards could see a different future for the land, one that would heal the land with a special kind of agriculture modeled on nature's patterns. They could imagine how the landscape could evolve into something that was not a farm or a forest. Something completely new, yet rooted in the ancient wisdom of the place.

Drawing inspiration from native ecological patterns common in the region prior to European colonization, the Shepard family began to carefully place trees, shrubs, vines, canes, grasses, forbs and fungi throughout the 106-acre farm to create healthy plant communities designed to produce food, fuel, medicines, and beauty. Because they needed to produce income while waiting for the perennials to produce marketable products, the farm design also included areas of annual crops like vegetables, hay, small grains and pastured livestock. "We got started by selecting perennial plants that mimicked the oak savanna plant community that we could sell, feed to an animal or eat ourselves," Mark recalls. "As things have matured through the years, we can afford to do less and less annual cropping. The products that we actually sell haven't changed much over the years, but the proportions of each have changed through time."

Credit: Restoration Agriculture Design

Mark Shepard

Today, New Forest Farm is a nationally recognized model for the successful transformation of an industrial grain operation into a commercial-scale, locally-adapted, perennial agriculture system. Hazelnuts, chestnuts, walnuts, apples and elderberries are the primary woody crops on the farm. In the alleys between a diverse mix of trees and shrubs, livestock—cows, pigs, turkeys, sheep, pigs or chickens—graze pastures of mixed fescues, clovers and wild plants grown in rotation with annual vegetables. The farm's principal products supply regional and national wholesale markets through the Organic Valley cooperative and the American Hazelnut Company. Small volumes of a diverse line of fresh and locally processed fruit, nut and livestock products are sold in local direct markets. The farm has been certified organic since 1995, is entirely solar and wind-powered, and farm equipment can be powered with locally-produced biofuels.

The innovative nature of New Forest Farm's production system design has influenced Mark's view of long-term production challenges. Most difficult starting out was the lack of marketable perennial fruit and nut varieties adapted to growing conditions in his region. "For example, chestnut and hazelnut varieties simply did not exist for this location," Mark explains. "Cold hardiness and disease tolerance with the chestnuts had a lot to do with it. With the hazelnut, everything had to do with the fact that the current industry is built around the European hazelnut, which will not survive here." One of the handful of early adopters to cross European and American hazelnuts, Mark has worked to create new cultivars that could thrive in his region. "What we're finding out is that the European genetics confer very few advantages to us. Through time, our plants are beginning to look more and more Americanoid. They're shrubs, not small trees. Cultivars that were my best plants 20 years ago are no longer my best plants. The genetic mix is constantly changing, and evolving and adapting." Mark is convinced that managing locally adapted cultivars for continual evolution is key to the continued success of his operation.

Mark anticipated long-term challenges associated with the well-known weather extremes typical of the region, so he put some careful thinking into water management. Working towards a goal of zero surface runoff, Mark modified keyline design principles to design and install earthworks that promote the distribution, retention and recycling of all the precipitation that falls on the farm. "Since we've been here," says Mark, "our longest drought was two calendar years where we had snow in wintertime, but almost zero measurable rain in the summer. Then in 2018 and 2019, we had twice the annual amounts of rainfall. Seventy-five inches of rain one year to zero inches of rain the next year. That's a challenge." Weather patterns in 2020 and 2021 have been uncharacteristically "normal," aside from warmer than usual weather in the fall. "Having predictable rainfall and tolerable temperatures has made me feel a little suspicious," Mark says, "and left me wondering when's the hammer going to fall, but I'll take every stretch of favorable weather we can get!"

Marketing the diversity of crops produced at New Forest Farm represents another long-term production challenge, because of the difficulties of managing diverse product lines in the current regulatory environment. "Managing a system like this, you've got to make sure that there's enough income coming off of each stream in order to support the infrastructure that it requires to get it to market. That's challenging, but worst of all are the regulations. At one point in time, we had six different USDA regulations governing our enterprises and each one of them had a different inspector that went with it. Complying with all these regulations in a diverse system is borderline insanity." For example, marketing organic produce requires USDA organic and GAP certifications. Wholesale and retail sales of processed products like meats, hard cider and hazelnut butter require food processing, meat retailing and winery licenses.

Other long-term challenges include variable spring weather that disrupts pollination and fruit set in spring-blooming fruit trees like apples and meeting the farm's need for seasonal labor. "We are in an agricultural area," says Mark, "so we have a lot of agricultural families around. Even so, to find seasonal labor when we need it has been just a royal nuisance."

Although weather variability and extremes are to be expected in the Midwest climate, Mark thinks that variability in rainfall and temperature have grown more intense in the last decade or so. "This part of the country has wide swings in temperatures, like a summertime high of 110 degrees and a wintertime low of 50 below zero," Mark explains. "That's a challenge for the basic selection of varieties that will work, but that's not the problem. The real challenge is to go from one day at 85 degrees, beautiful weather, leaves are emerging, and then have it instantly turn into rain, turn into freezing rain, turn into 18 inches of snow, clears off and then goes to ten below zero. That's brutal. That happened one Friday the 13th when I had just received several new honeybee colonies. It did not go well for them. The rapidity of temperature changes, especially in the springtime, is just phenomenal. I don't recall it being that extreme in the earlier years. That's been a noticeable change."

These changing weather patterns accumulate through the year to disrupt the production of fruits and nuts at New Forest Farm. "Temperature and precipitation variability and extremes, all of that, it all piles on," says Mark. "With the tree fruits in the past five years, I've had three of those years that for all practical purposes, zero apples. This is an apple-growing region. That's crazy. So probably five or six years out of the last 25, weather has totally short-circuited apple production in a historically apple-growing region. This is one of the reasons why monocrops seem like insanity to me. During the years when I had no apples, at least I had other enterprises that still yielded well. Several apple orchards went out of business while others relied on government bailouts. One orchard that had invested in an apple pie processing line was able to survive by relying on frozen inventory to stay afloat. I was able to rely on other income streams."

Working with hazelnuts in the variable Midwest climate, Mark is impressed by their robust nature. "I've seen the American shrub hazelnuts do absolutely everything," says Mark. "I've seen the temperature drop to five below zero when they're in full bloom, singes off all the female flowers and they just make more female flowers. I've seen them thrashed by hail, shattered to the ground by windstorms, tornadoes, I've had cattle browse them down to the nubbin. Those American bush hybrid hazelnuts are almost bombproof." Other species that stand out as resilient to variable weather conditions at New Forest Farm are elderberries, currants and the perennial pasture that makes up the orchard floor. "It doesn't matter whether it hasn't had rain in two years or not," says Mark, "the pasture will grow, cattle will eat it and you can sell a steer."

The diversity of species and products that Mark manages reduces the risks of weather variability to the whole farm, but individual species are still sensitive. For example, extreme weather variability in spring reduces yields of spring-flowering fruits like apples, but does not affect later-flowering nut crops. On the other hand, continuous rainfall throughout the 2019 growing season caused a 70 percent crop loss on chestnuts. "I've been selecting for years for the varieties that do really, really well here," says Mark, "but that doesn't change the basic fact that chestnuts don't like wet feet. We had the wettest two years on record here—two years in a row—and we have heavy clay soils."

The mix of annual and perennial species at New Forest Farm gives Mark some flexibility to respond to weather extremes that are unavailable to producers focused on one or the other. "If the annual crop fields are too wet to get into, I don't go into them," explains Mark. "For example, in the summer of 2019, I did zero annual crops. I just couldn't get into the field and even if I had been able to get into the field to plant, there's no way I would have been able to do any weed control. Annual crop farmers were stuck, they got hosed that year. Whereas I had other crops that were doing okay, just forget about the squash and peppers."

Thinking over all the fruit species he manages, Mark points to three that are challenging to grow at New Forest Farm. "Apricots, apples and chestnuts are the three most persnickety," says Mark. "One of the reasons why the chestnuts are challenging is that Chinese chestnuts theoretically won't survive in this zone. I expected 99 percent of my initial planting of Chinese chestnuts to die and they did. I also knew the American chestnuts I planted would get chestnut blight and they have. Those were the knowns. The hybrid crosses between the two, some of them got the blight, some of them didn't, some of them can handle the cold, some of them can't, but I keep on bringing in genetics from wherever I can find them and throwing them in the mix. If they fail to thrive, that's fine. I don't care about the ones that don't survive. I'm looking for the ones that actually thrive."

A key to Mark's success in developing locally-adapted fruit and nut cultivars is selecting for hyperprecocity—a characteristic that shortens the time it takes for a perennial plant to begin producing fruit. "People say, 'Well, it takes forever for a tree to reproduce itself,'" says Mark, "And I tell them 'Well, then those are the wrong trees.' We breed for hyperprecocity. I have two family lines of chestnuts that flower right out of the seed. We can cross-pollinate those and rapidly build locally adapted lines from there." Mark has also created a line of hazelnuts that flower in two years. "That's how we can turn around a new variety line so quickly. We can grab a plant from somewhere else, bring it in, have it pollinate our really fast hazelnuts, and then we've got the traits from those plants in our hazelnut breeding line. Even if that plant dies for whatever reason, that's fine. We've got its genetics or at least the important genetics."

Mark has not made any major changes in management to adapt to changing weather patterns, but he says he is always adjusting management practices based on his observation of New Forest Farm ecosystem processes. "As I learn, I do things differently than I used to do," says Mark. "That's the biggest thing." The evolution of the way he uses a subsoiler to capture and store precipitation in the soil is a good example. "In the early years, I would use the subsoiler to cut slots in the soil parallel to my tree rows in the early spring just after the soil thaws so that when the spring thunderstorms showed up, all that water would go down in the slots and migrate through the fields."

Two years in a row of extreme drought changed Mark's thinking. "The first of the two years in a row that I had zero rain, I went out and I used the subsoiler at the typical time. Well, no rain came. If you cut slots in the ground with no rain, you actually dry the soil out worse and faster. So I went and I used the subsoiler a second time that year, in the fall after everything was done for the season, with the thought that I'd be able to catch the water in the soil through the wintertime. That worked. That was what actually allowed me to harvest our annual crops that summer with zero precipitation because I'd captured all that melted snow and rain on frozen ground." Since then, Mark has switched to subsoiling in the fall.

Mark is quick to point out that his subsoiler is not a Yeomans plow—a tool that has captured the attention of a lot of regenerative farmers. Mark owns a subsoiler and Yoemans plow and has used both at New Forest Farm. "This is what's amazing about the whole mystique around the Yeomans plow subsoiler," says Mark. "I can't tell that it does anything different at all. They are both just hooks that you drag through the ground to cut through a hard pan in the soil, to lift and shatter it. The only difference that I can tell is that the Yeomans plow is more expensive."

Mark points to the resilience of the region's plant communities as his most important climate risk management strategy. "Nature has been here with these species for a long time—anywhere between six thousand and nine hundred million years," Mark explains. "That's what really matters. These plant communities belong here and they've been through it all. They've been through Ice Ages, they've been through global warming, they've been through megadrought, they've been through all of it. From my perspective, the best model I have for a system that can survive all the changes, whether it's short-term weather or long-term climate, is the plant community type that I select. Then all I've got to do is figure out how to dance with that going forward, making sure that I have the traits in my gene pool to survive whatever the insult is. That's what I have a little bit of authority over—the selection and breeding of those plants—which depends on my management skill, my own creativity, observation, decision making, experience, knowledge, etc."

Mark is upbeat about his ability to continuing farming into the future. As climate change disruptions grow more frequent, his nursery business has grown exponentially. "There are lots and lots of growers that are interested in 'reality-adapted' cultivars," Mark says, "whether they are organic or chemical. Our conventional horticultural crops are all genetic weaklings. Ninety-nine percent of our horticultural crops are sissies. They're bred to be raised in a controlled environment. The controlled environment is not reality. In order to control your environment, it costs. It costs labor and it costs infrastructure. It costs time and it costs money. It costs inputs. That's a cost that we don't need if we breed and select our cultivars to actually survive in reality. I look for the varieties and cultivars which survive with sheer total utter neglect. Okay? If I have to do anything to these plants in order for them to produce heavy crops pest and disease free, they're out. They die and they're out of the gene pool."

Mark sees promise in the design and cultivar development strategies that he has innovated at New Forest Farm. "Since I'm using the same tricks that nature has been using to survive," Mark says, "I feel that I have the best tools in the tool kit. The real issues that I see are not weather and climate or plants, they're human-induced laws, rules, regulations, attitudes, and technology. Those are the real challenges."

Mark Shepard is the CEO of Forest Agriculture Enterprises LLC, founder of Restoration Agriculture Development LLC and award-winning author of the book, *Restoration Agriculture* and a companion volume, *Water for Any Farm*. He is most widely known as the founder of New Forest Farm, the first of its kind in the United States. Mark has developed and patented equipment and processes for the cultivation, harvesting and processing of forest-derived agricultural products for human foods and biofuels production. Mark serves on the board of directors of The Stewardship Network and is a farmer member of the Organic Valley cooperative. He teaches agroforestry, Permaculture, and Master Line water management worldwide.

Walker Miller
The Happy Berry, Six Mile, South Carolina

> I knew that frost was the biggest risk going into this. It's still the biggest risk, and it has gotten worse. In the 80s, we would typically start frost protection in April. Now we start as early as the first week of March, so we're also at risk for a longer period of time, because we still can get a freeze through to the end of April.
>
> — Walker Miller

When Walker Miller was looking to start a fruit farm, he knew that one of the biggest risks he would have to navigate would be late spring frost. He also knew that kudzu could help him find some frost-protected land, because kudzu flourished in the warmer places in the mountains of South Carolina. He was looking for the perfect place to grow fruit, a place just above the colder bottom land as his first line of defense against late spring frosts.

Forty-two years ago, Walker found what he was looking for. "We have mountains to the north and west of the farm," Walker explains, "with Lake Keowee in between, and a ring of hills around the farm. When the cold air slides off the mountains into the Keowee River valley and settles on Lake Keowee, the warmer air on the surface of the lake is pushed up and over our farm." And so Walker and his wife Ann got started bringing new life to an old worn-out cotton hill farm, farming at night and on weekends when they weren't busy with their day jobs working in agricultural research and extension at Clemson University. Until her death in 2021, Ann and Walker managed the farm with the help of their daughters, Betty Ann and Zoe, a few seasonal workers and volunteers.

The Happy Berry Farm produces blackberries, blueberries, seedless grapes, muscadines, seedless muscadines, figs, persimmons and pussy willows, plus a number of minor crops such as mulberries, olives, chestnuts, and tea.[1] The farm totals about 22 acres of steeply sloped, highly eroded and erodible land that was farmed for cotton starting in the early 1800s and then abandoned from about 1930 until Walker and Ann purchased the farm. Market production is focused on about 14 acres, with about three acres in infrastructure support land, parking, driveways and buildings. "From the get go, marketing the farm was a key part of our plan," Walker says. "We wanted to focus on the pick-your-own market with wholesale as a secondary, so finding a location that was surrounded by five medium-sized towns and one major metropolis was ideal."

Credit: The Happy Berry

Walker Miller

Crop management at The Happy Berry features sustainable agriculture practices that support the cultivation of perennial fruits in a permanent sod cover. Organic mulches, swales and bio-retention areas help to capture and manage rainfall, and use a mix of cultural, biological and plant pharmaceuticals to promote plant health. Farm products are marketed as pick-your own and customers are encouraged to "graze while picking, because there is no better way to learn what is a ripe berry." Anyone feeling guilty about how much tasting they did while picking is invited to make a contribution to the "Sin Bucket" near the checkout center. Pre-picked fruit can be purchased on-farm at the porch of an old tenant house that serves as farm headquarters and at several weekly farmers markets.

Thinking back over his 40 years of fruit growing at The Happy Berry, Walker identifies managing pests and diseases as his most difficult production challenge, partly because it is constantly changing. "Weeds, insects and diseases—the list is quite long," Walker says, "and invasive species are growing more challenging in this area as the climate has changed. The kudzu stink bug, the spotted wing drosophila and the brown marmorated stink bug are here. The spotted lantern fly and others are headed our way. When I look at my weed spectrum, over 50 percent of my weeds are invasive species. Robins are also a perennial pest for us. They arrive in late July and leave in late August. For the first time in 2020, we were visited by blackbird flocks. I don't know what that portends." Walker uses Bird Gard speaker systems to warn migratory flocks that the farm is not a safe stopping place.

The farm's degraded soils and lack of water are two additional long-standing production challenges. Typical of many old hill farms in the southeast U.S., nineteenth-century farming practices stripped the hillsides of topsoil and dumped it into local waterways to be carried out to the Atlantic Ocean, leaving behind a mosaic of sandy and clayey acidic subsoils colonized predominantly by deciduous trees and kudzu after abandonment. When Walker purchased the farm, soil organic matter levels were about 0.5 percent, but 40 years of practices designed to restore healthy soils have raised soil organic matter levels to as much as nine percent in some parts of the farm, which helps to moderate water challenges. "That's one resource that I did not consider adequately when I purchased the farm," Walker says. "At the time, city water was cheap and so I opted to use city water. I've since put in two wells, both of which were a bust. Water continues to be a challenge for us."

Completing seasonal field work in a timely manner at The Happy Berry has always been a challenge because both Ann and Walker worked full time off the farm. Changing weather patterns have added to this challenge over the last decade. "Some of the windows for field operations are closing down," Walker explains. "They have gotten shorter, so I've got less time to complete the work. For example, I don't like to start winter pruning blueberries until I have accumulated 600 chilling hours, but it's taking longer and longer to reach that target because winter temperatures have warmed and become more variable."

Based on his farm records, Walker estimates the total number of chill hours has declined from about 1,400 hours to about 1,100 hours since he started growing at The Happy Berry. Any unseasonably warm periods in winter reverse the chill hour total and extend the time required to meet crop chill requirements. "Having to wait until January to start pruning presents a problem, because it takes a long time to prune. It takes three months just to do the blueberries. So with that window closing, I'm forced into a situation where I have to hire more help. The shorter the window, the more difficult it gets. You're trying to get pretty complicated tasks done and you have less time to do them. So it's just really adding to the challenges."

Other changing weather patterns that have complicated fruit production at The Happy Berry include more variable spring weather, high temperatures and heat waves in late summer, more heavy rainfalls and longer dry periods and droughts. "I knew that frost was the biggest risk going into this," Walker says. "It's still the biggest risk, and it has gotten worse. In the 80s, we would typically start frost protection in April. Now we start as early as the first week of March, so we're also at risk for a longer period of time, because we still can get a freeze through to the end of April." Plants are particularly sensitive to extremes of temperature during flowering and fruiting. For example, temperatures over 85 degrees can kill blackberry flowers, disrupt pollination and damage developing fruit. Long periods of days over 85 degrees are becoming more common in late summer and fall.

Walker thinks that seasonal patterns of rainfall are becoming more variable and weather extremes more frequent. "Our usual drought here is two to three weeks," Walker explains, "or it used to be, anyhow. Now we're getting droughts that can go ten weeks. Of course our response here is irrigation, but irrigation is very expensive for us, because we are on city water. And storms seem to be getting more violent, which presents additional problems. When you've got a bunch of blueberries on the bush and you get a really violent storm, it can knock a lot of berries to the ground."

To manage these new weather-related risks, Walker has made some changes at The Happy Berry. He has added a wind machine to enhance frost protection and he has dropped raspberries and elderberries from his crop mix as rising temperatures and more variable weather have reduced yields and increased insect and disease management challenges. On the plus side, a longer growing season and rising temperatures, especially in winter, have made it possible for him to add figs to his crop mix—a fruit that is in high demand in his area. Walker is also exploring the potential of adding chestnut and jujube[2] to his crop mix, because these late-blooming species are more likely to escape frost damage. As an added plus, they also add more diversity to his product line.

According to Walker, his long practice of selecting varieties for frost and disease tolerance continues to pay off as weather variability has increased. "I have a blueberry variety called Centurion that was developed in North Carolina. I have personally measured flower temperatures down to 17 degrees in that variety and had those flowers developed normally, which I think is absolutely amazing. I have grown Centurion for about 25 years and it has never missed a crop. Never." He has had similar success with Chester, an old USDA blackberry variety. Selecting blueberries that tolerate leaf disease has also been important "because disease pressures are increasing as our weather becomes more and more tropical."

Walker has been experimenting with a bit of ecosystem engineering as a strategy for moderating the risk of rising late-summer and fall temperatures in his blueberries and blackberries. "There is a commonly occurring association of pine with blueberry and ground-nesting bees in natural blueberry bogs," Walker explains. "So I decided to try planting some pines among my blueberry bushes for some shade. I planted them in east-west rows, with different spaces between the rows and the trees." Walker is testing three pine species—loblolly, Italian stone pine and longleaf pine—and he is also interested in the possibility of marketing the pine nuts produced by the trees.

According to Walker, pines have a number of characteristics that add to the utility of this polyculture as a climate resilience strategy. "First, the pines we've planted are non-epicormic, so when you cut a limb off, it does not grow back. This allows us to limb the trees up as they grow so they don't interfere with the wind machine when we need it for frost protection. Second, they provide passive frost protection by reducing heat radiation to the north sky. Third the needles and the roots are slower to break down. Fourth, mycorrhizae associated with pine roots can reach down as much as nine feet into the subsoil, retrieving phosphorous, nitrogen, sulfur and water in hydraulic redistribution that feeds both the pine and the blueberry. I believe there is a 'hand shake' between the mycorrhizae associated with the pine and the blueberry. We now have berry bushes which literally hug the pine trees. The only disadvantage is that the pines provide perching areas for migratory birds. There is still a lot to be learned, but I don't know of anybody out there doing this sort of thing, so I've just got to do it and see what happens."

Another new practice that helps to reduce climate risk at The Happy Berry is the use of cover crops to improve soil health. "As an example," Walker says, "in the blueberries we're mowing annual ryegrass in the middles with side-delivery mowers that blow everything underneath the bush to provide a soft carbon that decomposes quickly in the row. We drop the pruned branches in the middles and then use a rotary mower to chop and blow them underneath the plants to add some harder carbon that is slower to break down. We also blow all the pine needles underneath the plants. Our middles are in perennial summer grasses and we overseed them with the annual ryegrass to keep them biologically dynamic." In his grape plantings, Walker uses crimson clover cover crops to enhance soil health and provide a mulch to reduce weed pressure.

Walker has explored the potential to produce biochar on the farm to further enhance soil health. "You don't have to go very far around here to see pieces of charcoal where somebody has had a fire," Walker explains, "and everywhere you see that, you see these green spots. I've got this waste stream on the farm, over 3,500 blueberry bushes each producing five to ten stems a year that we can't chop with the mower. It's very dense wood that produces a beautiful charcoal. I've made it here. It would do great things for our soil like increase cation and anion exchange capacity, water infiltration rate and water holding capacity." Biochar also appeals to Walker because it turns a farm waste into a resource that also extends benefits far beyond the farm. "You know," says Walker, "'the world's on fire,' to quote Naomi Klein. Part of our mission here is to put that carbon back into the ground instead of back up in the sky."

Walker can't say he is confident about the future of fruit growing at The Happy Berry if weather patterns grow more variable and extreme as expected, but he is quick to point out that some of the biggest challenges to the sustainability of the farm are not actually on the farm. According to Walker, they are found in a national food policy that keeps prices low and promotes industrial growers, in a consumer population that has no connection with farming and in an increasingly globalized economy. "I'm a strong supporter of regional economies," says Walker. "Molly Scott Cato's book, *Bioregional Economy*, really got me thinking. She shares lot of good ideas about how we could redesign our economy and good arguments for why we should."

Although Walker sees the need for bigger changes in society, he still does what he can to guide The Happy Berry through these changing times. "I think about adaptation and mitigation every day," says Walker, "and dream about new climate resilience practices that are best suited to our site and its limitations. That's kind of the bottom line, there, isn't it?"

Walker Miller is a longtime leader in both civic and sustainable agricultural organizations in South Carolina and speaks regularly about the role of local agriculture as a climate change solution. He maintains an innovative program of on-farm research to assess the climate resilience of new fruit and nut species and innovative production practices and has collaborated in regional fruit research programs for many years. Walker was named the 2014 Farmer of the Year by the Carolina Farm Stewardship Association in recognition of his innovative contributions to sustainable agriculture in the Carolinas.

Table 16.1. Resilient Agriculture Fruit and Nut Producer Stories

Review this list to learn more about all the stories shared by fruit and nut growers that are featured in the 1st and 2nd editions of *Resilient Agriculture*. The stories marked with an asterisk are included in this volume. Updated stories from the 1st edition, and links to more information about all of the *Resilient Agriculture* producers can be found at realworldresilience.com

Farm or Ranch (Region)	Scale	P	A	T	Featured Resilience Behaviors
Tonnemaker Hill Farm (NW)	120	X	X		Add annuals, shifted to direct markets.
Fillmore Farms (SW)	230	X	X		Shift to organic, added cover crops and winter irrigation.
Ela Family Farms (SW)	100	X	X	X	Shift to direct markets, diversified fruit cultivars, added annuals and on farm processing, added frost protection and more water.
Shepherd Farms (MW)	300	X		X	Orchard renovation, shift to disease-tolerant cultivars, improve on-farm processing, add custom-shelling operation.
Red Fern Farm (MW)	15	X	X	X	Shift to perennial polyculture, integrate livestock/mob grazing, add irrigation/tree shelters.

Table 16.1. (cont'd.) Resilient Agriculture Fruit and Nut Producer Stories

Farm or Ranch (Region)	Scale	P	A	T	Featured Resilience Behaviors
New Forest Farm (MW)*	106			X	Shift to perennial polyculture, integrate annuals and livestock.
Almar Orchards and Cidery (MW)	300	X	X	X	Shift to organic, integrate livestock, add on-farm processing and retail store.
Bishop's Orchards (NE)	320	X	X		Increase field equipment, diversify perennial crops, add annual crops, shift to direct markets, add agrotourism and on-farm retail store, add on-farm processing (winery).
The Happy Berry (SE)*	22	X	X	X	Add frost protection, add cover crops, drop sensitive species, shift to frost and disease-tolerant cultivars, interplant shade trees, shift to less sensitive species.

Note: Scale is reported as acres under management; P, A, T refer to types of adaptive strategies used by the producer: Protect, Adapt and Transform (see Fig. 11.1), respectively.

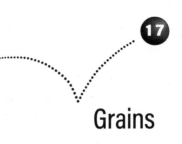

Grains

Bryce Lundberg
Lundberg Family Farms, Richvale, California

It just seems like we used to have a lot more regular storms that would come through. They weren't five or six inches of rain or ten feet of snow type storms, they were just regular, consistent rain patterns. Now it seems like there isn't such a thing. It's either really wet, or really dry.

— Bryce Lundberg

"'Leave the land better than you found it,' was one of those phrases we heard often when we were younger," says Bryce Lundberg, a member of the third generation of his family to produce, process and market rice in the Sacramento Valley of California. Bryce and his brother Eric, together with their wives Jill and Heidi, have grown rice on about 1,500 acres since 1985. Today, they are one of about 40 local farms who produce rice for Lundberg Family Farms on about 20,000 acres near Richvale.

Lundberg Family Farms supports a network of growers using organic and eco-farming practices that cultivate soil health, enhance biological diversity and reduce or eliminate the need to use synthetic fertilizers and pesticides.[1] These practices include crop rotation, cover cropping, innovative water management and soil incorporation of crop residues. Lundberg Family Farms growers also use practices that benefit wildlife, for example, by salvaging the eggs of waterfowl nesting on their farms in spring[2] and flooding their fields in winter to provide rich overwintering grounds for waterfowl.

These efforts to promote biodiversity extend to the diversity of rice varieties—17 at last count—currently produced by Lundberg Family Farms growers. "Some are easier to grow than others," Bryce explains. "We like to have the farms take a mix of varieties to spread the risk around of the hard varieties and the easier varieties. Some varieties just want to jump right out of the water, and other ones you really have to watch them a lot closer to make sure they're going to come out of the water." In any given year, Bryce grows a mix of red and black rice, Arborio, Jasmine and Basmati and also a variety of sushi rice called Calhikari. Bryce follows his rice crops with a winter cover crop mix of oats, vetch and fava beans.

Credit: Paolo Vescia Photography

Bryce Lundberg

Although managing water levels in the field to achieve goals like good crop establishment, reduced weed, pest and disease populations, and high quality grain is unique to rice production, seasonal patterns on a rice farm otherwise are typical of corn production throughout the U.S. Fields are prepared for planting in the spring, the crop flowers in late July and is ready for harvest in late September. Rice growers face many of the same longstanding production challenges managed by corn farmers everywhere, including weeds, seasonal water availability and field work delays caused by weather.

"Weeds are our big challenge," says Bryce. "Watergrass will just take over the field, and the weed seed is robust. You have to have a strategy to manage or control watergrass and the aquatic weeds like bulrush, the other sedges and the duck salad. If you don't, the rice is just not going to be financially viable. Fortunately over the years, we've developed some fairly effective strategies to address weeds."

Some of these strategies are unique to rice production, like managing weeds with flooding. For example, watergrass out-competes rice early in the season, so water levels are held high just long enough to drown the watergrass, but not so long that the rice drowns. It all comes down to the difference of just a few days. "The rice has about 48 hours longer that it can stay underwater than the grass," Bryce explains, "so that's your window for drowning grass." Bryce's dad and his uncles developed this innovative organic weed management strategy through careful observation of the interplay between water levels and crop and weed development in their rice fields.

Other longstanding management challenges at Lundberg Family Farms include seasonal water availability and weather patterns that delay spring planting. "Of course, we farm in California, so water is always a challenge," says Bryce. "Then late-maturing rice is always challenging, because we need to get it harvested before the rains come in the fall. Late-developing rice also often has low yields. Fortunately, we're in a dry and low-humidity growing region, so we don't have many problems with insects and disease. I would say the big issues for us have been weeds and late harvests."

Bryce sums up the weather changes that he has noticed over his years as a rice grower simply as fewer normal years. "We can have five years of drought in a row and then we can have floods," Bryce explains. "It just seems like we used to have a lot more regular storms that would come through. They weren't five or six inches of rain or ten feet of snow type storms, they were just regular, consistent rain patterns. Now it seems like there isn't such a thing. It's either really wet or really dry."

Along with more variable precipitation, Bryce has noticed warmer summer temperatures and more extreme weather. "The last five years we've had thunderstorms in the spring and fall with funnel clouds that have touched down around us," says Bryce. "The hail that comes with these storms can destroy a rice crop. I don't say we have Midwest weather here, but we have elements of Midwest weather that we never had. Tornadoes and hail are just not in our weather patterns here, but we've got a lot more of both. And that is a change. I can't tell you why."

More variable spring and fall weather have also challenged Bryce's ability to get field work done and complicates grain processing at Lundberg Family Farms. "The best rice is planted between the last week in April and May 15," Bryce explains, "so you have a short window in the spring when it isn't too wet or too cold to get your rice planted. Planting late is a problem. It pushes you late in the year and when you harvest rice late, it's poor quality. It's been rained on, it can get water stained, it can have a high moisture content. We like to harvest rice between 15 and 20 percent moisture and then dry it down to 14 percent in the granary. In 2016, when we got seven inches of rain during harvest, we were all in mud. There was not an easy acre of rice to harvest and we didn't finish harvesting until December. We were harvesting and receiving rice in our granary at over 30 percent moisture because if there was a day that fall that you could harvest, we just had to say, 'You know what, if you can harvest it, we'll take it.'"

In 2019, delayed planting in the spring had Bryce concerned about the fall harvest, but an unusually warm and dry fall created different kinds of challenges because the rice matured too quickly and all at once. "I would prefer last year (2019) to what happened in 2016," Bryce explains, "but it wasn't good for quality because it just dried up so fast. We had a three-week run where it didn't rain and it didn't cool off. It stayed hot and windy and just sucked the moisture out of the rice. Our growers just harvested rice faster than we could receive it at our granary. They couldn't cut it fast enough and we couldn't receive it fast enough."

Along with these specific weather-related challenges to crop production, the near failure of the Oroville dam in 2017 and a series of unprecedented wildfire seasons have left Bryce wondering about more general climate change-related threats to communities in his region. Many Lundberg Family Farms owners and growers were among the nearly 200,000 people living downstream of the Oroville dam who were abruptly evacuated in February 2017 after record-breaking winter rains damaged a spillway. And year after year of damaging wildfires have also left a lasting impression. "We've been a valley where we don't get the fires," says Bryce, "but the mountains are all around us and the fire seasons are awful."

Bryce has made a number of adjustments to his production practices to reduce the risks of more variable weather patterns. In order to have a bit more flexibility in planting and harvest dates, he has started substituting shorter-season rice varieties; however, a lack of diversity in commercial rice varieties, especially in short and long grain rice, limits Bryce's use of this strategy. "We don't have as many cultivars as I'd like to be able to choose between," Bryce says, "but there is some variation in season length." Short-season varieties also require less irrigation water, an additional plus as competition for water grows in California.

Another adjustment Bryce has made is to be better prepared to make the most of shifting field work windows in spring and fall. "When it's wetter, we're having to plant later, but really you can't always plant later," says Bryce. "The calendar is still the calendar. You've got a window of opportunity and so you have to work more hours, get more equipment and more people, and push them harder to get things done." More variable weather and shorter fieldwork windows also put additional pressure on the many agricultural service providers that rice growers depend on. "You can get the seed, you can get the fertilizer, but you can't always get it when you want it," Bryce explains. "When you have a more concentrated production season, everyone needs the service at the same time."

Bryce also manages risk through the purchase of government guaranteed production insurance, but he has rarely taken advantage of government programs that offer technical and financial assistance to implement agricultural conservation practices. "We do an awful lot of good management practices that could qualify, but a lot of times the rules are a little bit constraining," says Bryce. "There's a cost-share program that supports flooding for shorebirds. We like to flood for shorebirds, especially on the fields that are left to harvest the cover crop. But those rules are tough for us to manage. For example, one is to keep it wet a week or two longer than we want to. So we flood, but we won't sign up for the program, because we need more flexibility than the program allows."

Most recently, Bryce appreciated the assistance of local Resource Conservation District staff when he was ready to explore the potential of carbon farm planning. "I didn't know the first thing about writing a carbon farm plan," says Bryce, "and having those resources and that expertise come in from the government to help us understand how a carbon farm plan works on a rice farm...boy, was that helpful." Lundberg Family Farms has funded new research to explore greenhouse gas emissions in organic rice production systems in an effort to reduce emissions throughout their operations—both on the farm and in processing operations.

Lundberg Family Farms has also been a longtime supporter of Smartwater projects in California. "California's biggest water reservoir is the snowpack and if we lose it, that would be devastating," Bryce explains. "There's a reservoir proposed called Sites Reservoir that would capture and store flood flows from our snowpack. This is so important because with climate change, the idea is not that we're going to have less rain, it's going to be more rain, less snow, and so we're going to have more flooding. I think we've got to be prepared for that."

Thinking back over all of his experiences as a rice grower, Bryce appreciates the resilience benefits of his family's long experience working towards sustainability in organic rice production as part of a large network of growers. Over the years, this network has innovated organic and biodynamic rice production practices, wildlife-friendly water management practices that have helped to restore waterfowl and salmon populations in the Sacramento Valley, and water-efficient technologies like subsurface irrigation.

Bryce is upbeat about the future of rice production in the Sacramento Valley. "We're a resilient family and a resilient region," Bryce says. "We're confident that we'll keep farming rice. I see no reason we can't." He points to Lundberg Family Farms' newest acquisition, Dos Rios Farm, as a good example of what is possible when a business strives to "leave the land better than you found it."

Located at the confluence of the Sacramento and Feather rivers, the new farm is a prime site for agroecosystem restoration and management designed to benefit salmon and migratory birds, while simultaneously addressing flood management concerns and supporting local organic agriculture. "Farming inside the levee is a whole new experience for us," says Bryce. "The farm is located where the North State rivers come together and it's the first farm that floods.[3] It's very unique…we think it's the best farm in the state to grow salmon on in the winter time."

Bryce Lundberg serves as the Vice President of Agriculture, at Lundberg Family Farms, and is active in regional and state agricultural organizations, including the Northern California Water Association, the California State Board of Food and Agriculture, and California Certified Organic Farmers. Lundberg Family Farms, a producer of organic rice, rice products and US-grown quinoa, has long been widely recognized as a national leader in organic agriculture, most recently with the 2018 Leopold Conservation Award and the 2019 Organic Growers Summit Grower of the Year. In 2020, Lundberg Family Farms was recognized as an Organic Pioneer by the Rodale Institute.

Gail Fuller
Circle 7 by Fuller Farms, Severy, Kansas

We're still seeing the extremes. We've not experienced a huge drought since 2013. Actually, we've kind of seen the opposite. As of this June, we've gone 21 consecutive months with average or above average rainfall. And some months it wasn't just above, it was much above. In May 2019, 35 inches of rain fell on our farm— that's just one inch below the annual average, in one month.

— Gail Fuller

Gail Fuller owned and operated Fuller Farms, located in east central Kansas near Emporia for almost 30 years. In his effort to carry on his family's farming tradition, Gail transformed his 3,200-acre conventional corn and soybean operation into a 160-acre diversified pasture-based livestock farm producing nutrient-dense whole foods for direct markets. Gail's journey from Fuller Farm to Circle 7 is both a personal and a universal story of the barriers and opportunities that many American farmers are likely to encounter on the path to a resilient food future.

Gail first learned about farming from his grandfathers, who were both farmers, and then by working side by side with his father on their 700-acre family farm. In the late 1980s, Gail took over the grain production side of Fuller Farms. Like many producers in those times, he adopted no-till to try and reduce serious soil erosion problems and improve profitability. By the mid-1990s, he had dropped livestock from his operation and expanded corn and soybean production to more than 3,200 acres by leasing neighboring land.

Thinking back on the transition to no-till, Gail recalls following best management practices of the time which involved simplifying the farming system quite a bit. "During the '80s, we had a four-way rotation—corn, soybeans, wheat and milo—and we raised cattle. Everything except the corn and soybeans got kicked out with the big rush to no-till in the '90s. When no-till first really got popular, cows and no-till weren't allowed. It was thought at the time that cattle were too destructive to soils and the damage they caused by trampling farm ground couldn't be fixed without tillage, so the cows got kicked off."

Gail says that soil erosion did not seem to get much better with the switch to no-till, perhaps because the corn–soy rotation didn't leave much crop residue. "It was all corn and soybeans," Gail explains, "and most of the corn was being chopped for silage at the time, leaving very little crop residue in the fields. There was zero carbon in the system. The shift to no-till also created some issues with soil fertility and increased insect and disease problems."

By the late 1990s, Gail started adding cover crops into the rotation and brought cattle back to the farm to intensively graze pastures and cover crops in an effort to build soil quality and reduce soil erosion. Although early attempts to manage cover crops within the no-till system were challenging, diverse crop rotations integrated with cattle were a

Lynnette Miller
and Gail Fuller

central feature of Fuller Farms by 2003. Occasional crop nutrient or pest problems were easily managed in the well-established and highly diverse farming system.

For the next decade, Gail managed a large variety of cash crops, cover crops, cattle, sheep and poultry in a highly diversified and integrated dryland production system based on the principle of keeping a living root in the ground at all times. A typical cash crop mix in a given year might have included winter canola, winter barley, winter triticale, winter wheat, corn, grain, sorghum, soybeans, sunflowers, red clover, safflower, oats and peas. Cover crops added even more diversity on the farm, sometimes as much as 30 to 40 additional species. As biodiversity increased on the farm, Gail noticed that the fertility and pest challenges created by the shift to no-till were, as he puts it, "in the rearview mirror."

What was growing more challenging during those years was the weather. "While the '80s were challenging weather-wise," Gail recalls, "they don't hold a candle to what we've had here since 2000. Obviously, time can erase all those memories, but I don't remember the wild swings like we've seen over the last decade and definitely the '90s were not like this. Maybe on a grand scale we were starting to see these wild swings in the '80s and '90s, but they've just become much more defined and much

more sudden. Instead of having prolonged periods of below or above average, it seems like we just started going over cliffs all the time. Going
through the major drought of 2011, '12, and '13 and now the heavy rainfall
we've had since just confirms for me the importance of getting soil as
healthy as we can as fast as we can so that we can deal with what Mother
Nature's going to throw at us in the next ten years."

Along with more extreme drought and flooding rains, changing temperature patterns and especially warmer nighttime temperatures began
to reduce crop yields at Fuller Farms, according to Gail. "The winter
grains like cooler nights. They mature during late May and June and normally we're already getting pretty warm by then. We also get a lot of
humidity, so it's pretty hard to cool off at night." In 2012, temperatures
were so much warmer all through the spring and summer that everything was about 30 days early. The winter grains were stressed during the
grain-filling period by the hot, summer-like conditions in spring, while
corn and soybeans were stressed throughout the summer by excessive
heat. Gail recalls, "It was just over 100 degrees every day, day after day
that summer. If we could have cooled off at night and let those plants
relax a little bit, we probably would have had a better chance. It's really
the nighttime temperatures that got us more than anything. Obviously
110 degrees in the day is not anything to like, but when they can't cool off
at night it makes it so much tougher the next day."

Gail appreciated the flexibility his highly diversified production system gave him to fine tune crop management to seasonal weather conditions while also enhancing soil health and biodiversity on the farm. As
an early innovator of cover crop cocktails, Gail learned by doing how to
design species mixes to achieve multiple goals. "The first thing we do
is look at the resource concern we have in a particular field before we
design the mix for it," Gail explains. "Are you going to graze it? What
time of the year are you planting it? Then, if it looks like we will have a
certain kind of weather pattern, and we know that some cover crop species will handle that weather better than others, then we'll make them
the dominant species in the mix. And that's another thing about how a

mix increases resilience. When you guess wrong, there's still going to be something there that will grow. The more diversity we put in that mix, the more it protects us."

But resilience at Fuller Farms was not based just on crop diversity. Gail views the livestock as integral to the system as a whole. "The livestock became part of the crop rotation," Gail explains. "I had a twelve-month grazing plan. You have to have a plan if you're going to be grazing cropland. You need to have something in mind about where the cows are going to go in the worst-case scenario. In the future, the cows will be my crop insurance. That's how I plan to get income off of failed crops. I can graze them and then we will have added value to the crops and to the cattle. We eventually brought sheep into the operation, because they're more drought tolerant than cattle, so we even diversified there as well."

As Gail gained experience managing more diversified operations, he began to think about role of livestock on his farm in a different way. "Livestock became the driver," Gail recalls, "that got us moving towards becoming a perennial-based farm. We began to understand that we needed perennials to complete the water cycle, the carbon cycle, mineral cycle and to be able to harvest the maximum amount of energy possible." Inspired by the pasture-cropping practices[4] popularized by Colin Seis in Australia, Gail began to experiment with no-till planting of forages and grains into established warm- and cool-season pastures.[5] He also began to establish new pastures on his farm that mimic the grassland ecosystems typical of the region a thousand years ago. "We had some good early success with this," Gail says, "and it looked to me like we could make this work in just about any kind of pasture system."

But there was a downside to all this diversity. In 2012, Gail landed in the center of a national crop insurance controversy. When weather conditions made it impossible for him to kill a cover crop within the time required by insurance guidelines, his crop insurance was cancelled and his losses were not covered. The insurance company interpreted Gail's cover crop cocktails as intercropping—a prohibited practice.[6] In an ironic twist, Gail had previously been awarded a Conservation Innovation grant by the USDA's Natural Resources Conservation Service to pilot the new practice on the same fields deemed out of compliance by the crop insurance company.

After his request for a review of his case was denied, Gail took his concerns up the chain of command until he was invited to Washington, D.C., to discuss the situation with USDA agency directors and his congressional representatives. Gail remembers telling the group, "We've got three government agencies controlling production agriculture and none of them are on the same page, and that needs to stop. You need to pick a direction. We can't be getting paid here for one practice, then walk across the aisle and get denied benefits for doing the same thing."[7]

USDA agency representatives agreed and created an interagency task force soon after their meeting with Gail to try and harmonize cover crop policies. In April 2014, the task force released a new cover crop policy to be used by all three agencies, but the policy didn't help Gail. Even though he eventually won his battle with the USDA and received his 2012 insurance payout, the years of controversy cost Gail his line of credit at the bank, neighboring land owners cancelled land leases citing his "risky" farm practices and in 2014 the USDA denied yet another insurance claim.

In the aftermath, Gail decided to take a step back and reconsider his reason for farming. He and his partner Lynette spent some time looking at the kind of farming that would best help them reach their big-picture goals. "When I stopped and really thought about it, I realized that I was farming too much land. Trying to farm that many acres without any credit is impossible. I realized that I was just digging myself deeper into what was already a pretty deep hole. It wasn't long after that Lynette and I realized that downsizing the farm and getting more into direct marketing might be the solution. We didn't really set a goal for how small we wanted to get. I don't think we really knew, but by 2016 we were managing about 400 acres of my home farm."

As Gail and Lynette gained experience producing grain and livestock products for direct markets, a heavy debt load and increasing flood risk got them thinking about the potential benefits of selling the family farm and relocating to step down to an even smaller farm. "It wasn't just the finances, it was also a change in the weather," Gail explains. "The river had become so unpredictable. When you're trying to do pastured chickens and sheep and pigs in a small operation, living along a river isn't the smartest place to be. We had five consecutive weeks of flooding on our river the last year we were there. Prior to that, the worst flooding I ever experienced on our farm was in 1995 when our operations were disrupted for just a few days."

After years of continuing discussion, the Fuller family decided to sell the farm and in the fall of 2019 Gail and Lynnette moved to their new farm about an hour south of Emporia near Severy. Circle 7 by Fuller Farms is about 160 acres of diverse uplands that includes native grasslands, some annual croplands, seven ponds, an orchard and a purpose-built event center with a commercial kitchen. Although COVID delayed the launch of on-farm events in 2020, Gail and Lynnette had a successful first year on the new farm producing grass-finished beef and lamb, and pastured pork, chicken and eggs for direct markets. In 2021, they began to host a limited number of on-farm events.

Thinking about the future, Gail and Lynnette are excited by the potential to draw on everything they learned at Fuller Farms to achieve what Gail admits are "extremely lofty goals." They have plans to develop new vegetable and fruit enterprises to complement their livestock products, and Gail hopes to "bring the soils on this farm to the same capacity as the farm I left. It took me 20 years there, but I think we can do it here in five."

Gail Fuller's long history of innovative design and management has been documented in the USDA Natural Resources Conservation Service's Profiles in Soil Health and in Holistic Management International's Success Stories, and he was recognized by the American Soybean Association with a Conservation Legacy Award in 2013. He has hosted the popular Fuller Field School on his farm annually since 2012 and he is a founding board member of Great Plains Regeneration.

Table 17.1. Resilient Agriculture Grain Producer Stories

Review this list to learn more about all the stories shared by grain producers that are featured in the 1st and 2nd editions of *Resilient Agriculture*. The stories marked with an asterisk are included in this volume. Updated stories from the 1st edition, and links to more information about all of the *Resilient Agriculture* producers can be found at realworldresilience.com

Farm or Ranch (Region)	Scale	P	A	T	Featured Resilience Behaviors
Zenner Family Farm (NW)	2800	X	X		Increase crop diversity, add dynamic rotation, integrate livestock and intensive grazing of cover crops.
Lundberg Family Farms (SW)*	1500	X	X		Shift to shorter-season cultivars, increase field equipment, purchase production insurance, carbon farm planning.
Quinn Farm and Ranch (NGP)	4000	X	X		Drop sensitive crops, shift to fall-planted crops, continue to adjust crop rotation sequence, add cover crop cocktails and livestock, explore potential for no-till.
Brown's Ranch (NGP)	5000			X	Holistic management, shift to diversified rotations with livestock integration, grassland restoration with intensive grazing, livestock adaptation to local conditions, shift to direct markets.
Circle 7 by Fuller Farms (SGP)*	3200			X	Holistic management, shift to diversified rotations with livestock integration, add cover crop cocktails and livestock, shift to direct markets, downsize to 700 acres, retreat from floodplain and downsize to 160 acres at new location with agrotourism facilities.
Rosmann Family Farms (MW)	700			X	Diversified crop/livestock production, on-farm retail store, local foods café in nearby town.

Note: Scale is reported as acres under management; P, A, T refer to types of adaptive strategies used by the producer: Protect, Adapt and Transform (see Fig. 11.1), respectively.

Livestock

Albert Straus, Straus Dairy Farm
Marshall, California

This past year (July '20 to June '21) has been the driest year on record. California is currently experiencing extreme drought conditions throughout most of the state. It has never happened before that we've had so little rain and it looks like we're going to continue to have extreme shortages into the future.

— Albert Straus

Albert Straus returned to his family's dairy farm in the late 1970s during a time of dramatic change in the dairy industry. Small family dairies were under increasing pressure to either get big or get out of the business. Returning home after earning a Bachelor's degree in Dairy Science, Albert thought he could see a third option, one that put new best agricultural practices to work saving the family farm. In the early 1980s Albert and his father, Bill, implemented no-till seeding of crops to prevent soil erosion and reduce fuel consumption. They were already farming without herbicides or chemical fertilizers.

Despite these innovations, falling milk prices continued to threaten the economic viability of their dairy business and Albert wondered if the growing consumer interest in organic food offered a solution. He began to imagine a new kind of market for his milk, one that reflected the true costs of production, promoted responsible land stewardship and offered a viable, principled and sustainable business model for small dairy farms. Albert realized that going organic would allow him to fully embrace his deeply held belief in land stewardship, while also addressing the challenging economic realities of family farms in an era of intense industrialization.

Inspired by these ethical and economic considerations, Albert transitioned his farm to organic production in 1993 and founded Straus Family Creamery in 1994. "How do you create a viable farming system?" Albert asks. "That's the challenge we've tried to address with certified organic production and collaboration with the 12 family farms supplying our creamery. What I've tried to do is create a sustainable organic farming model that is good for the Earth, the soil, the animals and the people working on these farms, plus helps revitalize rural communities."

As founder and CEO of Straus Family Creamery, Albert strives to create a value network that respects and supports farmers by inviting them to inform company decision making. "To promote knowledge and relationship building between our employees and the farmers," Albert explains, "our team hand delivers milk checks twice a month to our farmers. Four times a year, we get together with all the farmers supplying the creamery to review our sales, decide on milk volume and price targets and talk about the challenges and the future of family farms. The whole idea is to create a sustainable pricing system for organic dairy farms that rewards the farmers for producing a high-quality product as well as all these other benefits to society."

Albert still manages the Straus Dairy Farm, getting up early every morning to oversee milking of the farm's 275 cows before heading to the office to manage daily creamery operations. The cows graze on 500 acres of pasture throughout the year (weather permitting) and are fed organic grains and forages such as triticale, alfalfa and okara (a soybean byproduct of tofu production). No antibiotics, hormones or GMO technology are used on the Straus Dairy Farm or any other farms supplying the creamery.

Thinking back over all the challenges he has faced as a dairy farmer, Albert points to the basic economics of farming in an industrializing food system as the most difficult. "We've gone from 4.6 million dairies in 1940 to less than 34,000 today," says Albert, "with almost nine percent of them going out of business in 2019. We're focused on this bigger picture and we are working to make a difference with a model of food and farming that is sustainable for the planet, the animals, the people and the community."

Managing herd health, particularly the health of younger calves, is a second long-standing challenge at the Straus Dairy. "Prevention is key," Albert explains, "because once you need to use an antibiotic or a hormone, the animal is no longer organic, and you have to remove them from the herd. Our farmers promote herd health through the use of certified organic practices that are widely recognized to cultivate soil health and enhance the well-being of the dairy herd." Pasture-based production, rotational grazing and the application of manure compost made on-farm are some key practices used by the farmers supplying milk to the creamery.

Credit: Straus Family Creamery

Albert Straus

The coastal climate and upland landscapes in Marin and Sonoma Counties enhance the management of pastured dairy operations. The area is not subject to intense extremes of hot or cold temperatures and flooding rains are rare, but the qualities of each year's rainy season—onset, length and total precipitation—are important factors in the success of farm operations. "The length of our growing season is determined by how much rainfall we have," Albert explains. "Lack of rainfall is probably the biggest weather-related challenge for us," says Albert. "We've never had much water here, so it's always been a precious resource."

Managing pastures as the rainy season has become more variable has been challenging for the creamery's farmers, because organic certification requires farmers to maximize the amount of grass consumed by cows throughout the year, no matter the weather. Although a management challenge, lengthening the grazing season not only enhances the health of the dairy herd, but it also helps to keep down feed costs, which are normally 40 to 60 percent of a dairy farm's gross income. "The one thing that I don't know if people understand is that pasture is the cheapest feed that you can feed the cows," Albert says, "so if you can maximize grazing, it helps tremendously. For example, during a normal rainy season, we can reduce our purchased feed costs by half, which is huge. The more pasture we have, the better we can do economically."

The critical role that grazing plays in organic milk production explains Albert's commitment to soil health practices. Compost application and soils with higher organic matter support a longer grazing season because the soil captures, stores and slow releases the moisture from the rain that falls on the farm. "We continuously improve our pastures to retain more moisture," Albert explains, "which allows more pasture growth. We also use rotational grazing to enhance soil and pasture health. We intensely graze an area, then move the cows and let the area rest for approximately 30 days until it grows again. We have enough different areas of pasture to be able to do that. Intensive grazing really helps improve the quality of the pasture as well as increasing the yield." The application of compost has also played an important role. "By adding compost to the land, we actually increased soil organic matter on our farm to six or seven percent says Albert. "For every percent of organic matter in the soil, you retain about an inch of water."

More frequent and intense weather extremes in recent years have been challenging, both on the farm and at the creamery. "Probably over the last decade, I'd say, we've had new challenges with drought and more extreme weather. In 2020 and 2021, we have not had any significant rain at all. This past year (July '20 to June '21) has been the driest year on record. California is currently experiencing extreme drought conditions throughout most of the state. It has never happened before that we've had so little rain, and it looks like we're going to continue to have extreme shortages into the future." Other examples of recent weather-related disruptions include major flooding in forage production areas in the northern part of the state that made it difficult to source organic feed and more frequent power disruptions as the wildfire season has grown longer and more intense.

Although worrying, Albert believes that his best drought management strategy is to continue his focus on pasture management practices that promote soil and animal welfare. One way that he is improving pasture management in the creamery's farm network is to collaborate in carbon farming research designed to document the resilience benefits of rotational grazing and compost application on his farm. "We've worked with a NASA scientist to quantify the growth in the pastures over the last 20 years with satellite imagery," says Albert. "The results showed that organic dairy farmers could save a minimum of $100 per acre per year of avoided purchased feed costs with rotational grazing plus compost application. We were the first dairy farm to participate in the Marin Carbon Project's pilot carbon farm plan program."[1]

Another way that the creamery achieves its mission is to encourage collaborative research and data sharing among its farmer network. "Getting other points of view from peers or people in the industry always helps," says Albert. "Experience and knowledge sharing is really important." For example, as state and federal support for the organic dairy industry has declined, the creamery has collaborated within their farmer network to generate the critical data needed for the adaptive management of dairy operations. "The state of California used to do cost of production surveys for organic dairies," says Albert, "but they stopped. So we are helping our farmers get their own financial data and compare it against the average of the group of farmers. It's going to be an ongoing program where we can share ideas within the group, so each member can work to improve their business."

Albert is also leading research on carbon management at the Straus Dairy Farm for the network. "I am working towards my own farm being carbon neutral by 2023 through a combination of carbon farming and emissions reduction practices," Albert explains. "For example, we're participating in the first commercial trial to reduce enteric methane production by feeding red seaweed[2] to the cows, and we're part of the way through converting our farm vehicles to all electric. We are also keeping track of the farm's carbon footprint with the Cool Farm Tool."[3]

Thinking about the future, Albert sees both opportunity and challenge in the changing climate. "It's not going to be fun or easy, but it's going to be interesting, because climate change is going to put stress on the whole food system. I think it's going to be a lot more important to have local and regional food and farming systems. I think changing weather patterns and climate change are actually going to help us think more carefully about our values as a society."

He wonders if our society's current inability to look beyond our differences is one of the greatest barriers to acting on climate change. "Livestock vs. no livestock, natural land vs. managed land, this kind of extreme thinking is everywhere in our society," says Albert. "We're not working as a community and we're not listening to each other. We don't respect each other enough to work together to come up with a common vision and common strategy to move forward. I see this play out everywhere in life these days. We are going to have to come together. Otherwise, we're all going to go down together."

Albert Straus continues his family's legacy of environmentalism, sustainable practices and stewardship of the land by embracing the benefits of modern technology. The Straus Dairy Farm has transformed manure into renewable energy with biodigester technology since 2004, was the first dairy in California to create a carbon farm plan and is collaborating with BMW to pilot a "farm-to-electric" program that powers the automaker's electric vehicles. The Creamery, already a widely-respected leader in energy efficiency and zero waste processes, moved to a new facility in early 2021 to support more advanced sustainable practices. Albert's innovative leadership has been recognized with a 2017 Vision Leadership Award by the Specialty Food Association, he was named one of 2020's Most Creative People in Business by Fast Company and he joined the CLEEN Project as a federal advisory board member in 2021.

Jordan Settlage
Settlage and Settlage Farm, St. Mary's, Ohio

> Holy moly! In 2012, we had major drought which led us to buying irrigation equipment, because we had two million gallons of water stored in our lagoon that we could just stare at while our crops shriveled up and dried. And then 2015, same thing, super dry. Then we get a year like 2018/19, where we got rain from August of 2018 all the way until June of 2019. That's like ten months of just endless rain. And it was a disaster.
>
> — Jordan Settlage

Jordan Settlage has wanted to milk cows for as long as he can remember. Although dairying is part of his family's legacy, Jordan's grandfather got out of the dairy business in the early 1990s, one of many thousands of dairy farms forced out of business as the U.S. dairy sector industrialized.[4] Jordan's dad was happy to leave the demands of dairying behind to raise hogs and beef cattle instead. "I would tell my dad, 'When I grow up, I want to be a dairy farmer,'" Jordan recalls, "and he's like, 'That's hilarious. I grew up on a dairy farm, we're not milking cows.' In the fifth grade, I wrote the report about how when I grow up, I will be a grass-based dairy farmer. I still have that report."

With his father's blessing, Jordan worked at a neighboring dairy farm throughout his teen years. After graduating from high school, Jordan served for almost four years in the Army. He returned home in 2009 a combat veteran, ready to continue his education. "I graduated college in 2014," Jordan recalls, "and I was like, 'Hey dad, I still want to be a dairy farmer. I've been doing this for most of my life already. I want to

milk cows.' And so in the fall of '14, we started buying some equipment for cows and we started milking again." February 2021 marked Jordan's six-year anniversary milking cows.

Jordan is quick to point out that he didn't set out to be an organic farmer. His goal is to be a part of a healthy food system. "We're not organic to be organic," Jordan explains. "It's great that we're organic, we get paid more for what we're doing, but our goal is to create an entire system, an ecosystem, that's centered around life. It's a big circle. We want healthier soils and healthier animals that will end up as a healthier consumer." Jordan stays away from practices that, in his view, promote death, which he sees as the primary operational mode of conventional agriculture. "We try to keep away from all the 'cides' that are rampant in conventional agriculture: pesticides, insecticides, fungicides, herbicides, homicide and suicide." He goes on to explain, "The suicide rate is out of control in farming. Farmers are killing themselves at a rate like they were in the '80s,

Credit: Settlage and Settlage Farm

Jordan Settlage

and on top of that, you've got farmers shooting each other over dicamba drift.[5] How can we look at that and not see that maybe that's a problem?"

Over the past six years, Jordan has transformed his family's 500-acre farm into 450 acres of diverse perennial pastures that are home to a 200-cow dairy herd. The cows are outside on pasture nine to ten months a year, grazing on diverse legume and grass blends that change with the season. "We shoot for like a 70/30 blend on grass to legumes. Rye, meadow fescue, bluegrass, orchard grass, alfalfa, red and white clovers make up our cool season pastures. Eastern gamma grass and switchgrass, plus red and white clover make up the warm season perennials." The cows are also fed forage sorghum silage and haylage produced on the farm. The goal is to have the whole farm green and growing year-round, even in winter, so that the farm acts as a biological solar and water collector. "It's just a big sponge now," Jordan explains. "It's great. We're holding rain, we're catching everybody else's top soil that blows away when it's windy, and our water runoff is as clean as can be."

Jordan uses adaptive multi-paddock grazing, or AMP grazing, to manage for soil, pasture and livestock health by varying grazing time according to seasonal changes in weather. "We try to run 12-hour paddocks, but sometimes they stay in a paddock for 24 hours every once in a while," Jordan explains. "We manage a very aggressive rotational grazing system. The goal is to give the cows a fresh paddock after every milking. We're always trying to give our cows the cream of the crop, because we're pulling milk off of them. They're working hard ass." Jordan can be more creative with cows that are not in the milking herd, because their nutritional needs are less demanding. "We'll use them for mob grazing on some areas," Jordan says, "and we'll use them for leader/follower grazing sometimes and then sometimes we'll let them be on good pasture. You know, come April and May, we just don't have any crappy pasture. When it comes July, sometimes you just have that pasture that hasn't gotten a cutting yet, and that is amazing feed for heifers and dry cows."

Thinking back over the time he has been managing the farm, Jordan points to weather as the most challenging risk management factor. Other aspects of the farm, such as labor, cash flow, flies, weeds, marketing, etc., have been difficult to manage at one time or another, but many of these challenges have become "just a non-issue" with the shift to pasture-based production, plus cooperative marketing through Organic Valley—a national, farmer-owned, cooperative that produces, processes and markets organic produce and dairy products under the Organic Valley brand. It is the nation's largest farmer-owned organic cooperative and one of the world's largest organic consumer brands. Jordan does wonder about putting all his eggs into one marketing basket, so to speak. "We're on a contract with Organic Valley," Jordan says, "and that's working for us, but if that were to fall through then marketing would be back at the number one challenge. Especially with milk, because you can't just sell milk to people."

Number one on Jordan's list of weather challenges is an increase in the variability of precipitation and extremes. "Holy moly!" Jordan exclaims. "In 2012, we had major drought which led us to buying irrigation equipment, because we had two million gallons of water stored in our lagoon that we could just stare at while our crops shriveled up and dried. And then 2015, same thing, super dry. Then we get a year like 2018/19, where we got rain from August of 2018 all the way until June of 2019. That's like ten months of just endless rain. And it was a disaster. In spring 2019 we got these monsoon-type rains that stranded our cows out in the fields because there was so much water. We had to walk the cows back through water that was belly deep. And once it quit raining in June, then it just quit raining for about five or six weeks."

Temperatures also seem be more variable as well and the growing season has lengthened, according to Jordan's father. Temperatures are warming, but extremes of both hot and cold temperatures are also becoming more common. "In 2014 and '15, we had incredibly cold winters," Jordan says. "From 2002 to 2014 we'd never had anything like it where you're having pipes freezing in the barn. Then boom, next winter, same exact thing. And then this winter (2019) was super, super mild, and yet we had incredibly cold weather the first week of November. We've also had more wild temperature swings, where it will be hot and we'll open all the barns up to get good fresh air and then the next day, it's freezing, windy, raining. I don't want to say I'm sold or not sold on any of this, but this weather is not normal. Whether it's global warming or cooling or just climate change, either way it is not normal."

As the weather has become more variable and extreme, Jordan appreciates the resilience benefits of the diverse perennial forages that are the foundation of his operation. "We have such diverse pastures," Jordan explains, "because we want to have stuff in there that will do good if it's hot, if it's cold, if it's wet and if it's dry. Not necessarily great, but good. We're trying to hit all those ranges." Other practices that have enhanced the resilience of the farm include the use of diverse cover crop mixes in the forage crop rotation, planting pollinator strips to encourage beneficial insects and trees for windbreaks in their setback acres, and following the cows with poultry in the pasture rotation to spread manure and reduce cattle pests, particularly flies.

Jordan also credits his family's ability to be creative while solving specific challenges with solutions that are life-centered. "We try to be creative," Jordan says. "Or we're just strange people, I guess you could say. We just try to attack our problems head on and look for a solution that is life-centered. For example, we raise our calves on the cow, because we did a really bad job of raising them on the bottle. We started letting the moms raise them and it's like, Oh my gosh, they do a great job. They are cleaning them all the time, they play with them, the calves exhibit all sorts of interesting behaviors that we've never seen in calves before and their health is amazing."

Other resources that have been important to climate risk management at Settlage and Settlage Farm include local technical assistance and new forage-making technologies. Local USDA Farm Service Agency and Natural Resources Conservation Service (NRCS) staff offered helpful technical planning and cost-share assistance for cattle lanes, fences, and water lines as Jordan transitioned the farm from annual row crop production to AMP grazing and he appreciated the opportunity to enroll a hundred acres in the grassland conservation program offered by the NRCS. Jordan also finds inspiration in the growing network of young farmers in his area, most of them organic. "Those are the farmers that I like to hang out with. Instead of just complaining about how things are like most of the older farmers do, we talk about how to make things better. I'd say we're all on the more positive side."

New technology imported from Ireland has also been an important key to Jordan's success as a grazier. "We pay for custom bailing," Jordan explains, "and they use a wrapped bailer developed in Ireland where they've dealt with wet weather for a lot longer than we have. It just poops out a bale that's done, you don't have to do anything else to it, which is super helpful for our risk management when it comes to weather variability, because you can knock hay down in the morning and bale it that afternoon. Then it can sit in the field and get rained on and it's fine. So that's massive for us."

Jordan doesn't pay much attention to weather forecasts, except during haymaking season. He decided to stop a daily check of the weather amidst the 2012 drought because he found the forecasts too discouraging. "Since then, it's been really freeing," says Jordan. "I just don't pay attention to it. It's not that I don't care. Every day I just know I'm going to wake up, the sun's going to rise, I'm going to milk the cows, and I will deal with what today brings me."

Jordan does appreciate having access to weather data to look up past precipitation and temperatures as a way to check his perception during periods when the weather seems unusual. "That's how I figured out that 2015 was hot," Jordan explains. "What really got me wondering was a long stretch that summer with nighttime temperatures that never dropped below 68 degrees. And I would think, Am I losing my mind? What's going on here? It seems like we used to get nights in the 60s, and every night that year it's in the 70s. So I checked the data and it turns out that the nights *were* hot. That year nights were even in the 80s a few times. That is really hot for cows. Nighttime temperatures like that mean that they just never get a break from summer heat. If summer temperatures continue to rise, we'll have to put more effort into figuring out some kind of way to keep them cool in July and August."

Although Jordan does not know what new weather challenges might be ahead, he has faith in his ability to deal with whatever new weather challenges might come his way. "Once we see certain things have happened a couple of times, a couple of months in a row, we just think, OK, this is a new thing that is happening, how can we handle it? We'll just figure out how to deal with it as it comes."

Thinking more broadly about the future, Jordan wonders about the potential for agriculture to be part of the solution to climate change. "I get tired of hearing other farmers saying stuff like 'The government needs to get away from what we're doing, because the farmers know best,'" Jordan explains. "Well, if we know so much, why does the Mississippi Delta look the way it does? I don't know if this is true, but I have heard that the carbon lost from soils due to poor farming practices is partly responsible for global warming. And it's like, hmm, yeah, one culprit for that is farmers. I guess I think there are too many people maybe in total denial that we have any impact on it, right? We know that if soil organic matter in the country goes up by one or two percent, there'd be a lot of carbon put back in the ground. And soils would hold a lot more water and keep it on the farm, which in the end would help the farmers too."

Jordan Settlage serves as a master grazier in the Ohio Ecological Food and Farming Association's Dairy Grazing Apprenticeship program, is a featured Organic Valley dairy farmer, leads a California Certified Organic Farmer consumer education project, and regularly posts dispatches from the farm on his YouTube channel, "Jordan Settlage."

Jamie Ager
Hickory Nut Gap Meats, Fairview, North Carolina

There are so many variabilities in farming that you can get all stressed out. Part of being a successful farmer is probably just your head space as it relates to these things. But the fact that we're having more unpredictable weather creates a low level of constant worry that can be taxing on the spirit.

— Jamie Ager

Amy and Jamie Ager and their three children, Cyrus, Nolin and Levi, are the fourth and fifth generations to grow food at Hickory Nut Gap, a 600-acre farm located on an old droving road in the Southern Appalachian Mountains just southeast of Asheville, North Carolina. Like many mountain farms in the region, Hickory Nut Gap has been home to a diversity of enterprises over more than a century of operation. Growing up on the farm, Jamie helped his family milk cows and raise beef cattle, hogs, poultry and apples.

Despite his parents' efforts to steer him away from a life of farming, Jamie had his eye on the family's old dairy barns as he thought about his future. "The farm was needing a new thing," says Jamie. "I saw an opportunity to differentiate ourselves and be able to make a living here." Just over 20 years ago and fresh out of college, Jamie and his wife Amy transitioned the farm to a rotational grazing operation and founded Hickory Nut Gap Meats.

In those early years, Jamie focused on farm management while Amy worked to develop regional wholesale and retail markets for their pastured meat products. When they could no longer meet the growing demand for meats produced on their own farm, Jamie reached out to other livestock farmers in the region to help. Today, Hickory Nut Gap Meats supplies a diverse line of pasture-raised products to local and regional, direct and wholesale markets centered in the Southern Appalachians and supports a network of more than 30 family farms growing 100 percent grass-finished beef and pasture-raised pork in North and South Carolina, Georgia, Virginia, Tennessee and Kentucky.

As one of the family farmers supplying Hickory Nut Gap Meats, Jamie oversees management of the farm's 120-head beef cattle herd, and raises 250–300 hogs, 3,500 broiler chickens, 600 laying hens and 350 turkeys annually on 200 acres of their home farm in Fairview, plus 200 acres of leased farmland. Both farms are certified organic. Perennial hay and forage crops grown on the farm are custom harvested. Conventional feed is purchased for the pork and poultry operations, but the farm is moving towards sourcing a local supply of non-GMO feed in 2022. Hickory Nut

Gap Meats operates a popular on-farm store, butcher shop and café and hosts a busy schedule of public and private events during the growing season. The Fairview farm is also home to a commercial heirloom apple orchard, an organic vegetable farm and a summer farm camp.

In 2008, Jamie's extended family placed the farm into a unique conservation easement managed by the Southern Appalachian Highlands Conservancy that allows for continuing agricultural use and the development of sustainable homesites on a parcel within a forested nature reserve.[6] "It would have really torn the family up to develop this place," Jamie recalled. "It was a tough family conversation for a number of years." Though Jamie and Amy Ager had already spent years building the business and improving the property, they couldn't fully commit to expansion until the easement was official. "That gave us the assurance that this is always going to be in agriculture and allowed us to start really investing in improvements to build opportunity here for future generations."[7]

Credit: Bren Photography

Jamie Ager

Thinking back over his time managing the farm, Jamie says that financial viability has been his biggest management challenge. "I would say that financial viability is the challenge that we spend the most time thinking about. Tied to the financial piece is that pasture-based producers haven't had an alternative market to sell to, so we had to build all of our sales and marketing side while also doing the production. That's just hard to do well, because you're so spread out. As we thought about the type of agriculture we wanted to promote, we realized that decoupling ourselves from commodity markets was a key. That's certainly what we spent most of our time focused on."

A second long-standing challenge that is also related to marketing has been access to USDA-inspected processing. "That's certainly an issue," says Jamie. "Processing is its own whole field of study, one that we continue to learn about and one that is continually changing. Processing logistics in the Southeast can be tough, because we have limited options at our scale of production and because it is so important to product quality. We pay a lot of attention to our processing plant partnerships."

According to Jamie, other production challenges at Hickory Nut Gap are all "just a part of farming." Weeds in pastures and field borders, livestock impacts on soil health, and managing water supply and fencing in the rotational grazing system stand out, but in general haven't created long-term issues on the farm. "In general, I would say that if you manage rotational grazing well, annual and perennial forage production takes care of most of these problems," Jamie says. "Weeds are certainly an issue, especially multiflora rose in this part of the country. But weeds are funny because we've had different weed issues over the years. We've had to just get better at adapting to changes in weed pressure. We're certified organic now, so we can't spot spray and stuff like that, which a lot of farmers can do with pastures. Weeds are always something we pay attention to, but they aren't a huge problem for us."

Take multiflora rose, for example. When Jamie set up the rotational grazing system at Hickory Nut Gap Farm 20 years ago, water quality regulations required him to fence all the creeks running through the farm to keep out livestock. "We were required to keep our cows out of the creeks because they were viewed as the source of the [water quality] problem, but really it is a management problem. So we did fence the cows out and all our creeks became a bunch of multiflora rose and weeds. It was a mess! So we decided to try managing the creeks with our cattle by flash grazing those areas [moving cattle very swiftly through an area] and it works. We have a much cleaner, healthier farm ecosystem now. People put environmental regulations in place with good intentions and in general they're a good idea, but inevitably regulations have some flaws in them. So it's always a balance, right? That's an interesting conversation."

Variability in precipitation has presented the greatest weather-related risk over the 20 years Jamie has been managing livestock at Hickory Nut Gap Farm, a risk that he thinks is growing. "Certainly the variability has become more of something that we pay attention to. Drought is something we've always had to deal with, but gosh, when it rains five inches, hurricane-style rains in May, it gets really difficult to operate in that kind of mud. So that's something I think has gotten more extreme. Keeping the pigs and cattle happy in long periods of soggy wet weather is probably our biggest management concern. And those are getting worse."

More generally, Jamie views dealing with weather variability as just a normal part of farming. "I used to get more worried about it," Jamie says. "There are so many variabilities in farming that you can get all stressed out. Part of being a successful farmer is probably just your head space as it relates to these things. But the fact that we're having more unpredictable weather creates a low level of constant worry that can be taxing on the spirit."

Although more variable precipitation has increased management stress on the farm, Jamie has not made any major adjustments in production practices. He points to the healthy pastures created by rotational grazing as his number one climate risk management asset. "We've got a heavy rain coming today and I know it'll be fine," Jamie explains. "It's good for it. The pasture grasses hold the soil, and they'll bounce back once you get a good sunny day behind it. Our pastures are not totally bullet-proof because of drought, but they grow. A well-managed pasture grows and you're always going to have grass in a rotationally grazed system. We've been okay when it gets dry and I really think that has to do with our grazing management over the last 20 years. We leave a lot of residual grasses when we graze, especially if it gets super dry. We don't graze it down to the nub, you leave some good residual and then the grass has a chance to use every bit of that dew in the morning. Certainly droughts stink and it feels good to get some rain afterwards, but in a system where you don't have that kind of resilience built in, it's much more stressful."

Jamie also appreciates the resilience benefits of participating in the regional network of pasture-based livestock producers that supply Hickory Nut Gap Meats. "Ultimately, it is the pastured-production models that drive the resilience of our farmers. As we build our network and support good management, the cream rises to the top. We host an annual meeting, present ideas on how we can do better. We are a community of producers working together to solve these problems. We don't have all the answers, but just the fact that we're having that conversation while we're sitting around eating lunch creates new partnerships that allow us to try out new things."

For example, it was a lunchtime conversation at the annual producers' meeting a few years ago that led to an innovative solution to the problem of finding a way to graze cattle through the winter. "Beef farmers in the mountains and Piedmont areas of North Carolina need a place to graze cattle in winter," Jamie explains, "and the eastern part of the state has a great climate to do that. Pork producers in eastern NC need a way to reduce the costs of adding winter cover crops to their grain rotation and offering winter grazing is a great way to do that. The pork producers also saw the potential soil health benefits of grazing cattle as an additional plus. So two of our producers decided to try it. And it worked. As we sat there looking at pictures of cows deep in lush cover crops in December at our producers' meeting the following year, I'm thinking, 'Home run, that's awesome.' The next winter, we sent 200 of our calves east to graze cover crops on our partner's pork farms."

As an additional benefit to their suppliers, in 2020 Hickory Nut Gap partnered with the Savory Institute to implement the Ecological Outcome Verification (EOV) protocol on their partner farms.[8] Farmers can use the EOV protocol to monitor how the land responds to their management decisions over time. The protocol tracks trends in biodiversity, soil health and ecosystem function (water cycle, mineral cycle, energy flow and community dynamics) in grassland environments, including natural and seeded grasslands, grazed orchards, silvopastures, integrated crop/

livestock operations and forested landscapes. The Savory Institute uses the EOV protocol to verify eligibility for participation in its Land to Market program[9] which connects conscientious buyers, brands and retailers directly to farms and ranches that are regenerating their land.

Thinking about the future, Jamie is confident that the farm's focus on healthy pastures will continue to be the key to success, especially if weather variability grows more intense as projected. "It seems like unpredictability is the name of the game, moving forward, right?" Jamie asks. "Nobody likes a hard driving rain, that doesn't help anybody. If that's what the climate models show are coming, then let's find ways to build farming systems to reduce those rains and the damage they cause. Perennial pasture is a great tool to build upon, along with healthy soil and water resources. When we think about the future here at Hickory Nut Gap, we ask ourselves, 'What are we good at?' Pastures are what we're good at, so we want to focus on that." He also views society's growing awareness of the need to act on climate change as an opportunity for pasture-based meat producers.

"We can see it as this big scary thing," Jamie explains, "but it's also an opportunity for pasture-based farmers to tell our story, talk to consumers, and to build a narrative around the different models of agriculture and how each drives ecosystem health, soil health, climate health and food resilience. Corporate agriculture is such a massive industry. They are so big that they can't be creative. They just see one way forward and that's the way they've always done things. I'm a big believer that we need to build the new better thing that's going to eventually replace the old thing. I think climate solution-style farming is the new thing. Our customers care about this stuff. There's a ton of investment in regenerative agriculture right now and so there's a ton of opportunity to help drive this change. We can get lost in the weeds worrying about it, or we can use the tools we have to create resilient farming systems. If we can make money on those systems and stay in business, then let's move. Let's do it."

Jamie and **Amy Ager** oversee the Hickory Nut Gap Meats team and growth strategy. The farm was designated a River Friendly Farm by North Carolina Soil and Water Conservation in 2001 and the North Carolina Farm Bureau recognized Jamie and Amy's innovative leadership with a Young Farmer and Rancher Achievement Award in 2011. In 2021, Hickory Nut Gap was featured on National Geographic Television's hit TV show, Gordon Ramsay: Uncharted.

Table 18.1. Resilient Agriculture Livestock Producer Stories

Review this list to learn more about all the stories shared by livestock producers that are featured in the 1st and 2nd editions of *Resilient Agriculture*. The stories marked with an asterisk are included in this volume. Updated stories from the 1st edition, and links to more information about all of the *Resilient Agriculture* producers can be found at realworldresilience.com.

Farm or Ranch (Region)	Scale	P	A	T	Featured Resilience Behaviors
Straus Family Creamery (SW)*	500	X	X	X	Shift to intensive grazing, add processing/retail marketing, growers' network, carbon farming.
CS Ranch (SW)	138,000	X	X		Holistic management, dynamic stocking, shift to no-till and to multi-use perennial forage species in irrigated pastures, add local foods café in nearby town.
Frasier Farms (SW)	29,000	X	X		Holistic management, dynamic stocking, cow-calf plus stocker operation, long-term weather forecasts, subsidized production insurance.
77 Ranch (SGP)	2500			X	Holistic management, reduce herd size, grasslands restoration with planned grazing, grazing cover crop cocktails, carbon farming.
Cates Family Farm (MW)	930	X	X	X	Restoration of oak savanna with intensive grazing, reduce herd size and increase rotation speed to leave more residual, add irrigation, expand slightly, increase stream crossings.
Gunthorp Farms (MW)	225			X	Shift to on-farm processing and direct marketing, multi-species pastured livestock, on-farm charcuterie.
Settlage & Settlage Farm (MW)*	500			X	Shift to intensive grazing grass-based dairy production.

Table 18.1. (cont'd.) Resilient Agriculture Livestock Producer Stories

Farm or Ranch (Region)	Scale	P	A	T	Featured Resilience Behaviors
Sap Bush Hollow Farm (NE)	160	X	X	X	Shift to intensive grazing multispecies pastured livestock production, direct markets, add backup solar, drainage, raised barn, ponds, reinforced poultry shelters, FAMACHA monitoring system, mob grazing, shifted lambing season.
Hickory Nut Gap Meats (SE)*	400			X	Shift to regenerative grazing multispecies pastured production, direct markets, growers' network.
Happy Cow Creamery (SE)	90			X	Shift to intensive grazing pasture-cropping "12 Aprils" production system, on-farm processing, on-farm store and local wholesale.
White Oak Pastures (SE)	2500			X	Shift to regenerative grazing multispecies pastured livestock production, add on-farm processing, direct marketing, agrotourism, solar farm, carbon farming.

Note: Scale is reported as acres under management; P, A, T refer to types of adaptive strategies used by the producer: Protect, Adapt and Transform (see Fig. 11.1), respectively.

Reading Guide available for download at:

https://newsociety.com/pages/resilient-agriculture-2-reading-guide

Notes

Chapter 1: Waking Up to Climate Change

1. I use the terms "industrial thinking" and "industrial culture" to refer to the beliefs, values and attitudes associated with European colonialism and imperialism which justified the exploitation of land, people and community to acquire the raw materials and labor needed for the production of goods and services. Industrial thinking seeks to extract wealth from nature through the use of efficiency, specialization, standardization, consolidation and control without regard for ecological health or social equity and rests on the foundational assumption that economic growth is the best measure of community well-being. In contrast, "sustainable thinking" refers to the beliefs, values and attitudes broadly associated with the concept that the goods can be produced to meet the needs of the present without compromising the ability of future generations to meet their own needs.

Chapter 2: Climate Change Is Changing the Weather

1. Adam Smith, "2020 U.S. billion-dollar weather and climate disasters in historical context," *Climate Watch Magazine*, NOAA, 2021.
2. Unless otherwise noted, the bulleted statements in this section are adapted from NOAA's National Centers for Environmental Information Billion Dollar Weather and Climate Disasters: Events. https://www.ncdc.noaa.gov/billions
3. "Deadly Heat Waves Are Becoming More Frequent In California," *Science Daily*, August 26, 2009.
4. Adam Nossiter, "Drought Saps the Southeast, and Its Farmers," *New York Times*, July 4, 2007.
5. M. Alex Johnson, "Floods create economic catastrophe in Midwest," *NBC News*, June 20, 2008.
6. Adam Smith, "2010–2019: A landmark decade of U.S. billion-dollar weather and climate disasters," *Climate Watch Magazine*, NOAA, 2020.
7. Saskia de Melker, "A Sour Season for Michigan's Cherry Farmers," *PBS News Hour*, Aug 16, 2012.

8. "It's Official: 2012 Drought Cost Taxpayers a Record 14 Billion," *Taxpayers for Common Sense*, 2013.

9. Theopolis Waters, "U.S. Drought In 2013 Hurts Cattle Ranchers With Dry, Poor Wheat Crop," *HuffPost*, March 16, 2013.

10. "Some Valley farmers to get zero water allocation," *ABC30 Action News*, February 22, 2014.

11. You can learn about these three forms of Indigenous agriculture in Chapter 12 or dig deeper into Indigenous foodways at cultivatingresilience.com

12. Redrawn from Young and Steffen, "The Earth System: Sustaining Planetary Life-Support Systems," in *Principles of Ecosystem Stewardship*, Carl Folke, et al., Springer, 2009.

13. Laurence Crane et al., *Introduction to Risk Management: Understanding Agricultural Risks. 2nd Edition*, Extension Risk Management Agency, USDA, 2013.

14. Laura Lengnick, "What Can Farmers Do About Climate Change" (Series), *Climate Column*, National Farmers Union, 2017.

15. *Conservation's Impact on the Farm Bottom Line*, Environmental Defense Fund, 2021.

16. *Turning Soils into Sponges: How Farmers Can Fight Floods and Droughts.* Union of Concerned Scientists, 2017.

17. Léopold Biardeau et al., "Soil Health and Carbon Sequestration in U.S. Croplands: A Policy Analysis," Goldman School of Public Policy, University of California, Berkeley, 2016. Bronson Griscom et al., "Natural Climate Solutions," *Proceedings of the National Academy of Sciences*, 2017. Keith Paustian et al., "Soil C Sequestration as a Biological Negative Emission Strategy," *Frontiers in Climate*, 2019.

Chapter 3: Understanding Exposure

1. *Climate Change and Agriculture in the United States: Effects and Adaptation*, USDA Bulletin 1935, Washington, DC, 2013.

2. Katherine Hayhoe, et al., "Our Changing Climate," Chapter 2, *Impacts, Risks, and Adaptation in the United States:* Fourth National Climate Assessment, Volume II 2018.

3. *Climate Change 2021: The Physical Science Basis.* 6th Assessment Report. IPCC. 2021.

4. The regional summaries included in this section were adapted from the 4th National Climate Assessment, 2018. Agricultural production estimates were calculated using 2017 National Agriculture Statistics Service Production Summaries for various crops. The summary data reported in the tables in this

section are adapted from the 2018 SARE Bulletin *Cultivating Climate Resilience on Farms and Ranches* and are the most up-to-date regional data available as of August 2021. Current weather history and expected changes in weather can be found in the NOAA State Climate Summaries and the National Climate Assessment Regional Reports.

5. California Agricultural Statistics Review 2017–2018. California Department of Food and Agriculture, no date.

6. Chelsea Harvey, "Climate Change has Helped Fuel a Megadrought in the Southwest," *Scientific American*, 2020.

Chapter 4: Understanding Sensitivity

1. Adaptation Resources for Agriculture, USDA Climate Hubs, 2017.

2. The 4th National Climate Assessment, Chapter 3: Water, 2018.

3. See Rex Dufour, *Biointensive Integrated Pest Management*, 2001, for a comprehensive discussion of ecological pest management principles and practices, including a detailed explanation of the pesticide treadmill.

Chapter 5: Understanding Adaptive Capacity

1. "Ecosystem Services" in "Climate Change Effects on U.S. Agricultural Production," Chapter 5 in *Climate Change and Agriculture in the United States: Effects and Adaptation*, USDA Bulletin 1935, Washington, DC, 2013.

2. Emergent properties are those qualities and behaviors that can only be observed and measured when the system is functioning well as a whole. For example, flight is an emergent property of an airplane that cannot be observed unless all the parts of the airplane are in place and operating properly and the plane has fuel and a competent pilot. Adapted from Jim Wilson, *Changing Agriculture: An Introduction to Systems Thinking 2nd ed.*, Simon & Schuster Australia, 1995.

3. Preston Sullivan, *Applying the Principles of Sustainable Farming*, National Sustainable Agriculture Information Service, 2003.

4. In contrast, crops in an industrial agroecosystem are usually supplied with imported synthetic nitrogen produced through an energy-intensive industrial process. Typically, about half of the imported nitrogen exits the agroecosystem into the atmosphere (where it is a powerful greenhouse gas) or into ground or surface waters (where it is a powerful water pollutant).

5. "Drivers and Constraints Affecting the Transition to Sustainable Farming Practices," Chapter 6 in *Toward Sustainable Agricultural Systems in the 21st Century*, The National Academies Press, Washington, DC, 2010.

6. "Climate Change Effects on U.S. Agricultural Production," Chapter 5 in *Climate Change and Agriculture in the United States: Effects and Adaptation*. USDA Technical Bulletin 1935, 2013.

7. Ibid.

8. Nadine Marshall et al., "Enhancing Adaptive Capacity," Chapter 15 in *Adapting Agriculture to Climate Change: Preparing Australian Agriculture, Forestry and Fisheries for the Future*, CSIRO Publishing, 2009.

9. *Farm and Farm Family Risk and Resilience Toolkit*, Cooperative Extension Service, University of Delaware, 2020.

10. Peter Howe and Anthony Leiserowitz, "Who remembers a hot summer or a cold winter? The asymmetric effect of beliefs about global warming on perceptions of local climate conditions in the U.S." *Global Environmental Change*, 2013.

11. Rachel Schattman et al., "Eyes on the Horizon: Temporal and Social Perspectives of Climate Risk and Agricultural Decision Making among Climate-Informed Farmers," *Society and Natural Resources*, 2021; David Lane et al., "Climate change and agriculture in New York and Pennsylvania: Risk perceptions, vulnerability and adaptation among farmers," *Renewable Agriculture and Food Systems*, 2018; Amber Mase et al., "Climate Change Beliefs, Risk Perceptions, and Adaptation Behavior Among Midwestern U.S. Crop Farmers," *Climate Risk Management*, 2017; Rachel Schattman, "Vermont Agricultural Resilience in a Changing Climate: An Investigation of Farmer Perceptions of Climate Change, Risk, and Adaptation," University of Vermont, 2016; Gabrielle Roesch-McNally "Agricultural transformations: climate change adaptation and farmer decision making," Iowa State University Digital Repository, 2016; Allison Chatrchyan et al., "Understanding the Views and Actions of U.S. Farmers Towards Climate Change," Cornell Institute for Climate Change and Agriculture, Research & Policy Brief, 2016; Caroline Stephens, "Raising Grain in Next Year Country: Dryland Farming and Adaptation in the Golden Triangle, Montana," University of Montana, 2015; John Jemison, Jr. et al., "How to Communicate with Farmers About Climate Change," *Journal of Agriculture, Food Systems and Community Development*, 2014; Amber Campbell et al., "Agricultural Producer Perceptions of Climate Change and Climate Education Needs for the Central Great Plains," University of Nebraska Public Policy Center, Paper 154, 2014; Roderick Rejesus et al., "U.S. Agricultural Producer Perceptions of Climate Change," *Journal of Agricultural and Applied Economics*, 2013.

12. "Adapting to Climate Change," Chapter 7 in *Climate Change and Agriculture in the United States: Effects and Adaptation*, USDA Technical Bulletin 1935, 2013.

13. Ibid.

Chapter 6: Managing Climate Risk: Adaptation Stories

1. Nadine Marshall et al., "Enhancing Adaptive Capacity," Chapter 15 in *Adapting Agriculture to Climate Change: Preparing Australian Agriculture, Forestry and Fisheries for the Future*, CSIRO Publishing, 2009.
2. *Adaptation Resources for Agriculture*, USDA Climate Hubs, 2017.

Chapter 7: A New Way to Think About Solutions

1. Jim Wilson explains how systems thinking can help us manage change in agriculture by: providing effective ways to avoid our natural tendencies to procrastinate (a stitch in time saves nine), and to get lost in details (can't see the forest for the trees); providing a discipline to overcome our natural inclination to solve existing problems in isolation; encouraging attention to both the interrelationships between different parts of a system as well as the parts themselves; cultivating greater awareness of the meaning and purpose of a system and how the context within which the system operates is shaped by cultural values and beliefs. Jim Wilson, *Changing Agriculture: An Introduction to Systems Thinking 2nd ed.*, Simon & Schuster Australia, 1995.
2. Brian Walker and David Salt. "Practicing Resilience in Different Ways," Chapter 5 in *Resilience Practice: Building Capacity to Absorb Disturbance and Maintain Function*, Island Press, Washington, DC, 2012.
3. A person, a farm, a community and an ecosystem are all examples of complex adaptive systems. Complex adaptive systems are defined by three requirements: (1) they are made up of independent and interacting components, for example, the species found on a farm; (2) the components are subject to forces that serve to promote or resist change in the system—for example, the population increase of a pest species and the predation of that pest by another species; (3) the system produces constant variation and novelty through changes in existing components or the entry of new components to the system, for example, the changes that accompany plant growth and development or the addition of new species in crop rotation.
4. "Understanding Agricultural Sustainability," Chapter 1 in *Toward Sustainable Agricultural Systems in the 21st Century*, National Research Council, The National Academies Press, Washington, DC, 2010.
5. Donald Nelson et al., "Adaptation to Environmental Change: Contributions of a Resilience Framework," *Annual Review of Environment and Resources*, 2007.
6. *Ten Principles for Building Resilience*, Urban Land Institute, 2018.
7. RELi—Resilient Design and Construction, U.S. Green Building Council, 2021.
8. Aaron King and Jared Green, *Professional Practice: Resilient Design*, American Society of Landscape Architects. 2021; The Sustainable SITES Initiative, Green Business Certification, Inc, 2021.

9. For example, Elena Bennet et al., "Toward a More Resilient Agriculture," *Solutions*, 2016; *The Climate Resilient Farming Program*, New York State Department of Agriculture and Markets, no date; "Building a Resilient Future in Food and Farming," National Sustainable Agriculture Coalition, 2020; Margiana Petersen-Rockney et al., "Narrow and Brittle or Broad and Nimble? Comparing Adaptive Capacity in Simplifying and Diversifying Farming Systems," *Frontiers in Sustainable Food Systems*, 2021; Elena Bennett et al., "Ecosystem services and the resilience of agricultural systems," *Advances in Ecological Research*, 2021.

10. Recall "Farms and Ranchers Are Ecosystems" in Chapter 5: Understanding Adaptive Capacity.

11. Laura Lengnick, "Cultivating Climate Resilience on Farms and Ranches," USDA Sustainable Agriculture Research and Education Program, 2018.

Chapter 8: The Qualities and Behaviors of Resilient Systems

1. This discussion of the qualities of resilient systems is adapted largely from Brian Walker and David Salt, "The Essence of Resilience Thinking," Chapter 1 in *Resilience Practice: Building Capacity to Absorb Disturbance and Maintain Function*, Island Press, 2012.

2. Recall the discussion of ecosystems in Chapter 5: Understanding Adaptive Capacity.

3. Sally Goerner et al., "Quantifying economic sustainability: Implications for free-enterprise theory, policy and practice." *Ecological Economics*, 2001.

4. For example, Lillian Borsa et al., "How can resilience prepare companies for environmental and social change?" *Resilience: A Journal of Strategy and Risk*, Price Waterhouse Cooper, 2014; Aron Cramer and Christine Diamente, *Resilient Business Strategies: Decisive Action for a Transformed World*, Business for Social Responsibility, 2021.

5. Adapted from Figure 4 in Brian Fath et al., "Measuring Regenerative Economics: 10 Principles and Measures Undergirding Systemic Economic Health," *Global Transitions*, 2019.

6. These asset classes are standard expressions of the resources available to support the well-being of a family, a business or a community. These five classes always appear in any system of asset management, but some systems break these five classes into a greater number, for example, social assets are sometimes split into social, cultural, political, etc.

7. Recall the importance of capital assets to adaptive capacity discussed in Chapter 5: Understanding Adaptive Capacity.

8. Recall that "identity" refers to the characteristic structure, function and purpose of a system.

9. Broadly speaking, Indigenous foodways can be organized into three types that are associated with specific biomes. Horticulturalists evolved in forests, pastoralists evolved in grasslands and sedentary agriculturalists evolved in large river valleys. See discussion of Indigenous foodways in Chapter 12.

10. Donella Meadows, "Leverage Points: Places to Intervene in a System," in *Thinking in Systems: A Primer*, 2008.

11. Ibid.

12. Ibid.

13. Fred Magdoff and Harold van Es, "The soil degradation spiral," Fig. 1.1 in *Building Soils for Better Crops*, USDA Sustainable Agriculture Research and Education Program, 2009.

14. Donella Meadows, "Leverage Points: Places to Intervene in a System," in *Thinking in Systems: A Primer*, 2008.

15. Ibid.

16. Ibid.

17. Sally Goerner et al., "Quantifying economic sustainability: Implications for free-enterprise theory, policy and practice," *Ecological Economics*, 2009.

18. For example, David Yaffe-Bellany and Michael Corkery, "Dumped Milk, Smashed Eggs, Plowed Vegetables: Food Waste of the Pandemic," *New York Times*, April 11, 2020; Jesse Newman, "Closed Because of the Coronavirus, Restaurants Clear Out Their Pantries," *The Wall Street Journal*, April 4, 2020.

19. For example, Dawn Thilmany et al., "Local Food Supply Chain Dynamics and Resilience during COVID-19," *Applied Economic Perspectives and Policy*, 2020; Tim Woods and Mahla Zare, "National Farm Market Impacts from COVID," *Economic and Policy Update*, Department of Agricultural Economics, University of Kentucky, 2021.

20. Georgia Gustin, "U.S. Taxpayers on the Hook for Insuring Farmers Against Growing Climate Risks," *Inside Climate News*, 2019. Julian Reyes and Emile Elias, "Spatio-temporal variation of crop loss in the United States from 2001 to 2016," *Environmental Research Letters*, 2019. Noah Diffenbaugh et al, "Historical warming has increased U.S. crops insurance losses," *Environmental Research Letters*, 2021. You can also go to cultivatingresilience.com to dig deeper into the growing public costs of agricultural subsidies.

21. Bryan Walsh, "10 Years After the Great Blackout, the Grid Is Stronger—but Vulnerable to Extreme Weather," *Time*, Aug. 13, 2013.

22. Camela Minoiu and Sanjay Sharma, "Financial Networks Key to Understanding Systemic Risk," *IMF Survey Magazine*, 2014.

23. William Fry et al., "The 2009 Late Blight Pandemic in Eastern USA," *The American Phytopathological Society*, 2009.

24. "A systemic resilience approach to dealing with Covid-19 and future shocks," OECD, 28 April, 2020.

25. Sally Goerner et al., "Quantifying economic sustainability: Implications for free-enterprise theory, policy and practice," *Ecological Economics*, 2009.

26. These system-wide failures can be explained by a weakening of balancing feedbacks and a strengthening of reinforcing feedbacks as systems grow larger and more complex as discussed in Sally Goerner et al., "Quantifying economic sustainability: Implications for free-enterprise theory, policy and practice," *Ecological Economics*, 2009.

27. Brian Walker and David Salt, "The Essence of Resilience Thinking," Chapter 1 in *Resilience Practice: Building Capacity to Absorb Disturbance and Maintain Function*, 2012.

28. See discussion of adaptive capacity as a component of vulnerability in Chapter 5: Understanding Adaptive Capacity.

29. For example, *The Resilient Supply Chain Benchmark Methodology Report*, The Association for Supply Chain Management, 2021; Marta Negri et al., "Integrating sustainability and resilience in the supply chain: A systematic literature review and a research agenda," *Business Strategy and the Environment*. 2021; "Value Chain Risk to Resilience Initiative: Climate Risk Integration Framework Memo," BSR, 2021; Alexandre Karl et al., "Supply chain resilience and key performance indicators: a systematic literature review," *Production*, 28, e20180020. 2018.

30. For example, Matteo Bizzotto et al., *Resilient Cities, Thriving Cities: The Evolution of Urban Resilience*, ICLEI—Local Governments for Sustainability, 2019; The Resilient Cities Network, 2021.

31. The Sustainable SITES Initiative, Green Business Certification, Inc. 2021.

32. For example, The Resilient Cities Network delivers urban resilience in partnership with a global community of cities and Chief Resilience Officers, resilientcitiesnetwork.org; The U.S. Climate Alliance is a bipartisan coalition of governors working to accelerate new and existing policies centered on equity, environmental justice and a just economic transition to reduce GHG pollution and build climate resilience at the state and federal levels, usclimatealliance.org; the U.S. National Climate Assessment regularly reports on climate change impacts and effective mitigation and adaptation options in the U.S., globalchange.gov

33. See, for example, Margiana Petersen-Rockney et al., "Narrow and Brittle or Broad and Nimble? Comparing Adaptive Capacity in Simplifying and Diversifying Farming Systems," *Frontiers in Sustainable Food Systems*, 2021; Andrea Basche et al., "Evaluating the Untapped Potential of U.S. Conservation Invest-

ments to Improve Soil and Environmental Health," *Frontiers in Sustainable Food Systems*, 2020; Laura Lengnick, *The Climate-Resilient Agriculture Initiative: Cultivating Climate Resilience in the Hudson Valley*, 2019; Laura Lengnick et al., "Metropolitan Foodsheds: A Resilient Response to the Climate Change Challenge?" *Journal of Environmental Studies and Sciences*, 2015; "Adapting to Climate Change," Chapter 7 in *Climate Change and Agriculture in the United States: Effects and Adaptation*, USDA Technical Bulletin 1935, 2013.

34. Resilience design criteria and associated agricultural practices are adapted from Joshua Cabell and Myles Oelofse, "An indicator framework for assessing agroecosystem resilience," *Ecology and Society*, 2012. Common farm sustainability indicators are adapted from Laura Lengnick and Susan Kask, "A Sustainable Decisions Tool for North Carolina Farmers," Prosperity Project, 2009.

Chapter 9: The Rules of Resilience

1. Rashawn Ray, "How We Rise: Five Things John Lewis Taught Us About Getting in 'Good Trouble,'" Brookings Institution, 2020.

2. I share three here: (1) Ubuntu is a Xhosa word originating from a South African philosophy that recognizes our common humanity, connectedness and interdependence, see *Everyday Ubuntu: Living Better Together, the African Way*, by N. Ngomane; (2) "the golden rule" is shared by the world's major religions; (3) this phrase is associated with the early years of the contemporary organic farming movement in the U.S.

3. For example, Oliver Curry et al., "Is It Good to Cooperate? Testing the Theory of Morality-as-Cooperation in 60 Societies," *Current Anthropology*, 2019.

4. This discussion is adapted from "Social Principles Essential to Economic Sustainability," in John Ikerd, *The Essentials of Economic Sustainability*, Kumarian Press, 2012.

5. Chayenne Polimédio, "Our Laser-Like Focus on Individualism Is Destroying Our Communities," *New America*, 2018.

6. Some examples of this conversation within the sustainable agriculture movement that informed my own resilience thinking: Brewster Kneen, *From Land to Mouth: Understanding the Food System*, University of Toronto Press, 1995; Jack Kloppenburg et al., "Coming in to the foodshed," *Agriculture and Human Values*, 1996; Patricia Allen, "Realizing justice in local food systems," *Cambridge Journal of Regions, Economy and Society*, 2010.

7. Allison Alkon and Julian Agyeman, "Introduction: The Food Movement as Polyculture," in *Cultivating Food Justice: Race, Class and Sustainability*, 2011; Rachel Slocum et al., "Solidarity, space and race toward geographies of agrifood justice," Justice Spatiale-Spatial Justice, Université Paris Ouest, 2016;

Elisabeth Farrell et al., "Equity as Common Cause: How a Sustainable Food System Network is Cultivating Commitment to Racial Justice," *Othering and Belonging*, 2017.

8. Robin Wall Kimmerer, "The Serviceberry: An Economy of Abundance," *Emergence Magazine*, 2021.

9. "Regenerative Agriculture: A decolonization and indigenization framework," Regenerative Agriculture Alliance, 2020; Rupa Marya and Raj Patel, *Inflamed: Deep Medicine and the Anatomy of Injustice*, Macmillan, 2021.

10. Mathias Wackernagel and Bert Byers, *Ecological Footprint: Managing Our Biocapacity Budget*, New Society Publishers, 2019.

11. "Global Footprint Network—Advancing the Science of Sustainability," Smart Economics for the Environment and Human Development, 2021.

12. *State of the States: A New Perspective on the Wealth of Our Nation*, Global Footprint Network, 2015.

13. "National Ecological Deficit/Reserves," Global Footprint Network, 2021.

14. Elena Bennet et al., "Toward a More Resilient Agriculture," *The Solutions Journal*, 2016.

15. Julie Kurtz et. al., "Mapping U.S. Food System Localization Potential: The Impact of Diet on Foodsheds," *Environmental Science & Technology*, 2020.

16. See discussion of resources under management in Chapter 5: Understanding Adaptive Capacity.

17. For example, "Community wealth building," Democracy Collaborative, 2019; Johanna Bozuwa and Thomas Hanna, "Building Community Wealth through Community Resilience," *Community Development Innovation Review*, 2019; "Measuring Rural Wealth Creation: A Guide for Regional Development Organizations," National Association of Development Organizations Research Foundation, 2016.

18. For example, Todd Schmidt et. al., "Measuring stocks of community wealth and their association with food systems efforts in rural and urban places," *Food Policy*, 2021; *Whole Measures for Community Food Systems: Values-Based Planning and Evaluation*, Community Food Security Coalition, 2009.

19. Brian Fath et. al., "Measuring Regenerative Economics: 10 Principles and Measures Undergirding Systemic Economic Health," *Global Transitions*, 2019.

20. For example, "Planning for Resilience with Frontline Communities," Regional Plan Association, 2021; "Community-Driven Resilience Planning: A Framework," National Association of Climate Resilience Planners, 2017; Chapter 10, "Agriculture and Rural Communities," Fourth National Climate Assessment, United States Global Change Research Program, 2018.

21. For example, Asheville Climate Justice Initiative, City of Asheville, North

Carolina, 2020; Climate Justice Alliance: Communities United for a Just Transition, 2021.

22. For example, "A Long Food Movement: Transforming Food Systems by 2045," The International Panel of Experts on Sustainable Food Systems, 2021; "Food Systems of the Future: A Synthesis of Global Reports," Meridian Institute/ Global Alliance for the Future of Food, 2020; "Incentivizing Food Systems Transformation," World Economic Forum, 2020; Walter Willett et al., "Food in the Anthropocene: the EAT–Lancet Commission on healthy diets from sustainable food systems," The Lancet Commission, 2019; Creating a Sustainable Food Future. World Resources Institute. 2018; The Future of Food and Agriculture (Series), UN-FAO, 2018.

23. The Future of Food and Agriculture (Series), UN Food and Agriculture Organization, 2021.

24. Daniele Giovannucci et al., "Food and Agriculture: The Future of Sustainability," UN Sustainable Development Program, 2012.

25. For example, see the comparison of the politics, discourse, production philosophy and policy solutions of four different food regimes (food enterprise, security, justice and sovereignty) in Eric Holt-Gimenez, "Food Security, Food Justice or Food Sovereignty?" This Food First Backgrounder (2010) for compelling examples of how sustainability thinking limits international discussions on the transformation of food and farming.

26. For example, Kees de Roest et al., "Specialization and economies of scale or diversification and economies of scope? Assessing different agricultural development pathways," *Journal of Rural Studies*, 2018; Michael Duffy, "Economies of Size in Production Agriculture," *Journal of Hunger & Environmental Nutrition*, 2009.

Chapter 10: Is Sustainable Agriculture a Resilient Agriculture?

1. Stephen Gliessman, "The Case for Fundamental Change in Agriculture," Chapter 1 in *Agroecology: The Ecology of Sustainable Food Systems Third Edition*, 2014; "Part 1: Outcomes," in *Introduction to the U.S. Food System: Public Health, Environment, and Equity*, 2015.

2. "Contending Philosophies of Agriculture," in Chapter 1,"Understanding Agricultural Sustainability," in *Toward Sustainable Agricultural Systems in the 21st Century*, The National Research Council, 2010.

3. Although no simple typology or set of categories can capture the complexity of the farming systems used in U.S. agriculture, I use the terms *sustainable* and *diversified* to refer to farming systems that emphasize the use of natural processes within the farming system, often called "ecological" or "ecosystem"

strategies, which build efficiency (and ideally resilience) through complementarities and synergies within fields, on the entire farm and at larger scales across the landscape and community. Such farming systems represent a major departure from the key features which characterize *industrial agriculture*: large size combined with a high degree of specialization, reliance on off-farm and synthetic inputs, and the production of commodities under contract to food processors and handlers. A *food system* is the complex set of actors, activities, and institutions that link food production to food consumption. Food systems differ from farming systems in that the primary focus is beyond the farm gate. This distinction is adapted from "A pivotal time in US agriculture," in *Toward Sustainable Agricultural Systems in the 21st Century*, 2010.

4. "The Case for Fundamental Change in Agriculture," Chapter 1 in Stephen Gliessman, *Agroecology: The Ecology of Sustainable Food Systems Third Edition*, CRC Press, 2014.

5. Charles Francis et al., "Food webs and food sovereignty: Research agenda for sustainability," *Journal of Agriculture, Food Systems and Community Development*, 2013.

6. Gigi Berardi et al., "Stability, Sustainability, and Catastrophe: Applying Resilience Thinking to U. S. Agriculture," *Human Ecology Review*, 2011.

7. Polly Ericksen, "What is the Vulnerability of a Food System to Global Environmental Change?" *Ecology and Society*, 2008.

8. "A Pivotal Time in U.S. Agriculture," in *Toward Sustainable Agricultural Systems in the 21st Century*, National Research Council, 2010.

9. John Beeby, "Climate Change and Grow Biointensive," *Ecology Action*, 2010.

10. ecoact.org

11. Jean-Martin Fortier, *The Market Gardener: A Successful Grower's Handbook for Small-Scale Organic Farming*, New Society Publishers, 2014; Andrew Mefferd, *The Organic No-Till Farming Revolution: High-Production Methods for Small-Scale Farmers*, New Society Publishers, 2019.

12. "Biointensive No-Till Farming Systems," Community Alliance with Family Farmers, 2021.

13. "Permaculture Solutions for Climate Change," permacultureclimatechange.org, 2021.

14. "What is Permaculture?" permacultureprinciples.com

15. Thomas Henfrey and Gil Penha-Lopes, *Permaculture and Climate Change Adaptation: Inspiring Ecological, Social, Economic and Cultural Responses for Resilience and Transformation*, Permanent Publications, 2016.

16. Wes Jackson, The Land Institute, landinstitute.org

17. The Savanna Institute is a nonprofit organization dedicated to developing restorative, savanna-based agricultural systems through research, education and outreach. savannainstitute.org

18. Carbon Harvest provides carbon planning and management services promoting carbon sequestration through edible agroforestry in rural, urban and suburban landscapes in the Southern Appalachian region. carbonharvest.co

19. "Biodynamic Principles and Practices," Biodynamic Association, no date.

20. "Climate Action Through Agriculture," Biodynamic Federation, 2021.

21. "Biodynamic Farm and Processing Standards," Demeter Association, Inc., 2021.

22. Peter Pearsall, "Farmer-Philosopher Fred Kirschenmann on Food and the Warming Future," *Yes Magazine*, 2013.

23. "Definition of Organic Agriculture," International Federation of Organic Agriculture Movements, 2020.

24. "National Organic Program," USDA Agricultural Marketing Service, no date.

25. Lisa Held, "The Real Climate Impact of Organic Farming," FoodPrint, Grace Communications Foundation, 2020.

26. For example, *Organic Agriculture in the Face of a Changing Climate—A Toolkit for Consumers, Advocates & Policymakers*, Organic Farming Research Foundation, 2020.

27. "Regenerative Organic Agriculture," Rodale Institute, 2021.

28. "Farm Like the World Depends on It," Regenerative Organic Certified, 2021.

29. "What is climate-smart agriculture?" CSA Guide, Consultative Group on International Agricultural Research, Research Program on Climate Change, Agriculture and Food Security, no date.

30. Sheri Spiegal et al., "Evaluating strategies for sustainable intensification of US agriculture through the Long-Term Agroecosystem Research Network," *Environmental Research Letters*, 2018.

31. Peter Lehner and Nathan Rosenberg, "Legal Pathways to Carbon-Neutral Agriculture," *Environmental Law Reporter*, 2017.

32. "Carbon Farming," Carbon Cycle Institute, 2021.

33. Polly Ericksen, "What is the Vulnerability of a Food System to Global Environmental Change?" *Ecology and Society*, 2008.

34. Greg Barnes, "Environmentalists continue battle with lawmakers, pork industry over biogas from hog waste," *North Carolina Health News*, May 2021.

35. Peter Newton et al., "What Is Regenerative Agriculture? A Review of Scholar and Practitioner Definitions Based on Processes and Outcomes," *Frontiers in Sustainable Food Systems*, 2020.

36. "Regenerative Agriculture," General Mills. 2021.

37. "A regenerative agriculture industry alliance for thriving businesses, people and planet," Regenerative Agriculture Alliance, 2021.

Chapter 11: Resilient Agriculture: New Tools for Shaping Change

1. David Salt and Brian Walker, *Resilience Practice: Building Capacity to Absorb Disturbance and Maintain Function*, Island Press, 2012.

2. See Ann Adams, *At Home with Holistic Management: Creating a Life of Meaning*, Allan Savory Center for Holistic Management, 1999, or John Gerber, *Create Your Personal Holistic Goal*, UMass Sustainable Food and Farming, for examples of adaptive management for individuals and families; Preston Sullivan, *Holistic Management: A Whole-Farm Decision Making Framework*, ATTRA, for examples of adaptive management for agricultural businesses; Keith Moore, "Landscape Systems Framework for Adaptive Management," Chapter 1 in *The Sciences and Art of Adaptive Management: Innovating for Sustainable Agriculture and Natural Resource Management*, Soil and Water Conservation Society, 2009; and for a general review of adaptive governance, Brian Chaffin et al., "A decade of adaptive governance scholarship: synthesis and future directions," *Ecology and Society*, 2014.

3. Keith Moore, "Landscape Systems Framework for Adaptive Management," Chapter 1 in *The Sciences and Art of Adaptive Management: Innovating for Sustainable Agriculture and Natural Resource Management*, Soil and Water Conservation Society, 2009.

4. "Managing Climate Risk: New Strategies for Novel Uncertainty," in Chapter 7, "Adapting to Climate Change," *Climate Change and Agriculture in the United States: Effects and Adaptation*, USDA Technical Bulletin 1935. 2013.

5. For example, Ronda Janke, "Whole Farm Planning for Economic and Environmental Sustainability," Kansas State University Agricultural Experiment Station and Cooperative Extension Service, Publication MF-2403, 2000.

6. For example, Alan Miller et al., "Farm Business Management for the 21st Century: Key Financial Performance Measures for Farm General Managers," Department of Agricultural Economics, Perdue University, ID-243, 2001.

7. For example, see Laura Lengnick and Susan Kask, "Helping Farmers Make Complex Choices: A Sustainable Decision Tool for Farmers," Lengnick and Kask. 2009; Elizabeth Henderson and Karl North, *Whole Farm Planning: Ecological Imperatives, Personal Values and Economics, 2nd Edition*, Northeast Organic Farming Association Interstate Council, Organic Principles and Practices Handbook Series, 2011.

8. Some common indicators recommended for use in monitoring farm sustainability include soil health, energy and water use, biodiversity, time for family, local sales, satisfaction from farming, total family income. Whole farm management guides typically include directions for the regular monitoring of farm performance using a suite of recommended environmental, social and economic indicators, for example, see the *Monitoring Toolbox*, Land Stewardship Project; the *Sustainable Decisions Toolbox for North Carolina Farmers*, Cultivating Resilience, LLC; and Elizabeth Henderson and Karl North, *Whole Farm Planning: Ecological Imperatives, Personal Values and Economics, 2nd Edition*, Northeast Organic Farming Association Interstate Council Organic Principles and Practices Handbook Series.

9. Jill Mackenzie and Loni Kemp, *Whole Farm Planning at Work: Success Stories of 10 Farms*, The Minnesota Project, Minnesota Institute for Sustainable Agriculture, 1999.

10. Ronda Janke, *Whole Farm Planning for Economic and Environmental Sustainability*, Kansas State University Agricultural Experiment Station and Cooperative Extension Service, Publication MF-2403, 2000.

11. For example, see *Organic Whole Farm Planning*, Ohio Ecological Food and Farming Association, 2005; *Whole Farm Planning for Beginners*, Virginia Cooperative Extension, no date; Farm Beginnings Whole Farm Planning Course, GrowNYC, 2021; Preston Sullivan, *Holistic Management: A Whole Farm Decision-Making Framework*, ATTRA-National Sustainable Agriculture Program, 2001.

12. "What is Holistic Management?" Savory Institute, no date.

13. "Why Holistic Management?" Holistic Management International, 2021.

14. Alexandre Chausson et al., "Mapping the effectiveness of nature-based solutions for climate change adaptation," *Global Change Biology*, 2020.

15. Andrea Basche, "Turning Soils into Sponges: How Farmers Can Fight Floods and Droughts," Union of Concerned Scientists, 2017.

16. Anika Terton and Julie Greenwalt, *Building Resilience With Nature: Maximizing Ecosystem-based Adaptation through National Adaptation Plan Processes*, International Institute for Sustainable Development, 2021; *An Ecosystem-Based Approach to Climate Resilient Agriculture*, EbA Info Brief #4, International Climate Initiative, 2021; Camilla Donatti et al., *Ecosystem-based Adaptation Essential for Achieving Sustainable Development Goals*, Conservation International, 2015; Hannah Reid et al., *Is Ecosystem-Based Adaptation Effective? Perceptions and Lessons Learned from 13 Project Sites*, IIED Research Report, 2019; *Ecosystem-based Adaptation and Food Security*, US Agency for International Development. 2017; *Ecosystem-based Adaptation and Extreme Events*, US Agency for International

Development, 2017; Guy Poppy et al., "Achieving food and environmental security: new approaches to close the gap," *Philosophical Transactions of the Royal Society B*, 2014; Richard Munang et al., "Climate Change and Ecosystem-based Adaptation: a new pragmatic approach to buffering climate change impacts," *Current Opinion in Environmental Sustainability*, 2013; Sara Scherr and Jeffery McNeely, "Biodiversity conservation and agricultural sustainability: towards a new paradigm of 'ecoagricultural' landscapes," *Philosophical Transactions of the Royal Society B*, 2008.

17. Jian Liu et al., *Restoring the natural foundation to sustain a green economy: A century-long journey for ecosystem management*, International Ecosystem Management Partnership, United Nations Environmental Program Policy Brief 6, 2011.

18. "Adapting Agriculture to Climate Change," in Chapter 7, *Climate Change and Agriculture in the United States: Effects and Adaptation*, USDA Bulletin 1935, 2013.

19. William Easterling, "Guidelines for adapting agriculture to climate change," in *Handbook of Climate Change and Agroecosystems: Impacts, Adaptation and Mitigation*, Imperial College Press, 2009.

20. Thomas Spies et al., "Climate change adaptation strategies for federal forests of the Pacific Northwest, USA: ecological, policy, and socio-economic perspectives," *Landscape Ecology*, 2010; *National Roadmap for Responding to Climate Change*. USDA Forest Service, 2010; *National Fish, Wildlife and Plants Climate Adaptation Strategy*, National Fish, Wildlife and Plants Climate Adaptation Partnership, 2012; Maria Janowiak et al., *Adaptation Resources for Agriculture: Responding to Climate Variability and Change in the Midwest and Northeast*. USDA Climate Hubs, 2016.

21. Maria Janowiak et al., *Adaptation Resources for Agriculture: Responding to Climate Variability and Change in the Midwest and Northeast*. USDA Climate Hubs, 2016.

22. "Understanding Agricultural Sustainability," Chapter 1 in *Toward Sustainable Agricultural Systems in the 21st Century*, National Research Council, 2010.

23. "Adapting Agriculture to Climate Change," Chapter 7 in *Climate Change and Agriculture in the United States: Effects and Adaptation*, USDA Bulletin 1935, 2013.

24. For example, in 2014 the federal government released the U.S. Climate Resilience Toolkit and USDA initiated the Climate Hubs program. The Climate Resilience Toolkit is an interactive website designed to serve interested citizens, communities, businesses, resource managers, planners and policy leaders at all levels of government by providing scientific tools, information and expertise to help people manage their climate-related risks and opportunities, and improve their resilience to extreme events. USDA's Climate Hubs

are designed to develop and deliver science-based, region-specific information and technologies, with USDA agencies and partners, to agricultural and natural resource managers to enable climate-informed decision making and effective adaptation.

25. Andrea Basche et al., "Evaluating the Untapped Potential of U.S. Conservation Investments to Improve Soil and Environmental Health," *Frontiers in Sustainable Food Systems*, 2020.

26. "Adapting Agriculture to Climate Change," Chapter 7 in *Climate Change and Agriculture in the United States: Effects and Adaptation*, USDA Bulletin 1935, 2013.

27. Ibid.

28. Ibid.

29. Roderick Rejesus et al., "U.S. Agricultural Producer Perceptions of Climate Change," *Journal of Agricultural and Applied Economics*, 2013.

30. "Adapting Agriculture to Climate Change," Chapter 7 in *Climate Change and Agriculture in the United States: Effects and Adaptation*, USDA Bulletin 1935. 2013.

31. Andrea Basche, *Turning Soils into Sponges: How Farmers Can Fight Floods and Droughts*, Union of Concerned Scientists, 2017.

32. "Adapting Agriculture to Climate Change," Chapter 7 in *Climate Change and Agriculture in the United States: Effects and Adaptation*, USDA Bulletin 1935, 2013; Giovanni Tamburini et al., "Agricultural diversification promotes multiple ecosystem services without compromising yield," *Science Advances*, 2020.

33. "Adapting Agriculture to Climate Change," Chapter 7 in *Climate Change and Agriculture in the United States: Effects and Adaptation*, USDA Bulletin 1935, 2013.

34. "Solutions," Project Drawdown, 2021.

35. Brian Walker and David Salt, *Resilience Practice: Building Capacity to Absorb Disturbance and Maintain Function*, Island Press, 2012.

36. "Adapting Agriculture to Climate Change," Chapter 7 in *Climate Change and Agriculture in the United States: Effects and Adaptation*, USDA Bulletin 1935, 2013.

Chapter 12: The Light and the Dark of These Times

1. William Rees, "The Human Nature of Unsustainability," Chapter 15 in *The Post Carbon Reader: Managing the 21st Century's Sustainability Crises*, 2011.

2. Resilience theory proposes that as disturbance offers opportunities for innovation or transformation, these opportunities can be managed to bring about more desirable systems; see Brian Walker et al., "Resilience, Adaptability and Transformability in Social-ecological Systems," *Ecology and Society*, 2004.

3. Michelle Miller et al., "Critical Research Needs for Successful Food Systems Adaptation to Climate Change," *Journal of Agriculture, Food Systems and Community Development*, 2013.

4. This discussion of pastoral, horticultural and sedentary agriculture foodways was adapted from Earnest Schusky, *Culture and Agriculture: An Ecological Introduction to Traditional and Modern Farming Systems*, Greenwood Press, 1989.

5. In 1987, the United Nations Brundtland Commission defined sustainability as "meeting the needs of the present without compromising the ability of future generations to meet their own needs."

6. For example, Wes Jackson, *New Roots for Agriculture*, University of Nebraska Press, 1980; *Sustainable Food Systems*, 1983; Joan Gussow and Kate Clancy, "Dietary Guidelines for Sustainability," *Journal of Nutrition Education*, 1986.

7. Jack Kloppenburg et al., "Coming Into the Foodshed," *Agriculture and Human Values*, 1996.

8. Brewster Kneen, *From Land to Mouth: Understanding the Food System*, 2nd Edition, 1995.

9. Brian Halweil and Thomas Prugh, "Homegrown: The Case for Local Food in a Global Market," Worldwatch, 2002.

10. According to the Regenerative Agriculture Alliance, decolonization is about restructuring the systems that currently are responsible for the destruction and degeneration of life on earth, while validating such actions in the name of progress and civilization. Decolonization is a process intended to transform ownership, control and governing structures currently responsible for the destruction of natural systems, expropriation of ancient native lands, the invalidation of farming, governing, and commons-based systems that promote a symbiotic relationship with nature, and the continued war against Indigenous-centered cultures. See "Regenerative Agriculture: A Decolonization and Indigenization Framework," Regenerative Agriculture Alliance, 2020.

11. For example, the *Real Food Challenge* was founded in response to growing movements for farmworker justice, international fair trade, student farms and gardens, and local food as a means to amplify student voices and focus collective efforts on real change in the food industry. The *National Good Food Network* promotes market-based solutions in food distribution that will bring more food from sustainable sources to more people and places. The American Planning Association was a key partner in *Growing Food Connections*, an effort to build local government capacity to enhance food security for all.

12. Christian Peters et al., "Foodshed analysis and its relevance to sustainability," *Renewable Agriculture and Food Systems*, 2009.

13. Holly Hill, "Food Miles: Background and Marketing," ATTRA, National Sustainable Agriculture Information Service, 2008.

14. John Ikerd, "Reclaiming the Heart and Soul of Organics," in *Sustaining People Through Agriculture Series*, 2008.

15. For example, Branden Born and Mark Purcell, "Avoiding the Local Trap Scale and Food Systems in Planning Research," *Journal of Planning Education and Research*, 2006; Josee Johnston et al., "Lost in the Supermarket: The Corporate-Organic Foodscape and the Struggle for Food Democracy," *Antipode*, 2009; Laura DeLind, "Are local food and the local food movement taking us where we want to go? Or are we hitching our wagons to the wrong stars?" *Agriculture and Human Values*, 2010; Allison Perrett and Charlie Jackson, "Beyond Efficiency—Reflections from the Field on the Future of the Local Food Movement," Vermont Food Systems Summit, 2013; Carrie Furman et al., "Growing food, growing a movement: climate adaptation and civic agriculture in the southeastern United States." *Agriculture and Human Values*, 2013.

16. Joan Gussow, "Dietary Guidelines for Sustainability: Twelve Years Later," *Journal of Nutrition Education*, 1999.

17. "Principles of a Healthy, Sustainable Food System," American Planning Association, 2010.

18. For example, Chelsea Canavan et al., "Sustainable food systems for optimal planetary health," *Transactions of The Royal Society of Tropical Medicine and Hygiene*, 2017.

19. Walter Willett et al., "Food in The Anthropocene: the EAT-*Lancet* Commission on Healthy Diets From Sustainable Food Systems," *The Lancet*, 2019.

20. Around the world, C40 Cities connects 97 of the world's greatest cities to take bold climate action, leading the way towards a healthier and more sustainable future. Building on the work of the Milan Urban Food Policy Pact, the C40 Food Systems Network supports city-wide efforts to create and implement integrated food policies that reduce greenhouse gas (GHG) emissions, increase resilience and deliver health outcomes. www.c40.org

21. For example, Joan Gussow and Kate Clancy, "Dietary Guidelines for Sustainability," *Journal of Nutrition Education*, 1986.

22. Charlotte Glennie and Alison Alkon, "Food Justice: Cultivating the Field," *Environmental Research Letters*, 2018; Leah Penniman, "4 Not-So-Easy-Ways to Dismantle Racism in the Food System, *Yes Magazine*, April 2017; Rachel Slocum et al., "Solidarity, space, and race: toward geographies of agrifood justice," *Spatial Justice*, 2016; Julian Agyeman and Jesse McEntee, "Moving the Field of Food Justice Forward Through the Lens of Urban Political Ecology," *Geography Compass*, 2014; Alfonso Morales, "Growing Food and Justice: Dismantling Racism Through Sustainable Food Systems," in *Cultivating Food Justice: Race, Class and Sustainability*, 2011; Patricia Allen, "Realizing Justice in Local Food Systems," *Cambridge Journal of Regional Economy and Society*, 2010.

8. Kate Clancy and Kathy Ruhf, "Is Local Enough? Some Arguments for Regional Food Systems," *Choices: The Magazine of Food, Farm, and Resource Issues*, 2010.

9. Ibid.

10. For example, Kathy Ruhf, "Regionalism: a New England recipe for a resilient food system," *Journal of Environmental Studies and Sciences*, 2015; Andrew Zumkehr and J. Elliot Campbell, "The potential for local croplands to meet US food demand," *Frontiers in Ecology and the Environment*, 2015; Julie Kurtz et al., "Mapping U.S. Food System Localization Potential: The Impact of Diet on Foodsheds," *Environmental Science and Technology*, 2020.

11. Diana Liverman and John Ingram, "Why Regions?" Chapter 13 in *Food Security and Global Environmental Change*, Routledge, 2010.

12. Angela Tagtow and Susan Roberts, Iowa Food Systems Council, 2011.

13. Center for an Agricultural Economy, 2011. Updated 2016.

14. Vermont Sustainable Jobs Fund, 2011. Revised 2021.

15. Delaware Valley Regional Planning Commission, 2011.

16. Food Solutions New England, 2014.

17. A project of the New England State Food System Planners Partnership, 2019.

18. Chesapeake Food Network, 2020.

19. Piedmont Triad Regional Food Council, 2021.

20. Greater Capital Area Food System Assessment, Capital Roots, 2021.

21. Brian Donahue et al., "A New England Food Vision," Food Solutions New England, 2014.

22. Ben Bowell et al., "New England food policy: building a sustainable food system," American Farmland Trust, Conservation Law Foundation, Northeast Sustainable Agriculture Working Group, 2014.

23. Kathy Ruhf, "Regionalism: a New England recipe for a resilient food system," *Journal of Environmental Studies and Sciences*, 2015.

24. "Symposium on American Food Resilience: Promoting a Secure Food Supply," *Journal of Environmental Studies and Sciences*, 2015.

25. Allison Blay-Palmer et al., "Validating the City Region Food System Approach: Enacting Inclusive, Transformational City Region Food Systems," *Sustainability*, 2018.

26. Steve Jennings et al., "Food in an urbanized world: the role of city region food systems in resilience and sustainable development," The International Sustainability Unit, The Prince of Wales Charitable Foundation. 2015; Blay-Palmer, Ibid.

27. Robert Hoppe and David Banker, "Structure and Finances of U.S. Farms: Family Farm Report 2010 Edition," USDA-ERS Economic Information Bulletin Number 66, 2010.

28. G. W. Stevenson and Rich Pirog, "Values-Based Supply Chains: Strategies for Agrifood Enterprises of the Middle," in *Food and the Mid-level Farm: Renewing an Agriculture of the Middle*, MIT Press, 2008; Michelle Miller et al., "Regional Food Logistics: A Stakeholder Process to Inform the Multisystem Redesign for Sustainability," presented to the National Transportation Forum, Atlanta, March 12–14, 2015.

29. Leslie Day Farnsworth and Michelle Miller, "Networking Across the Supply Chain: Transportation Innovations in Local and Regional Food Systems," Center for Integrated Agricultural Systems, University of Wisconsin-Madison, 2014.

30. Robert Hoppe and David Banker, "Structure and Finances of U.S. Farms: Family Farm Report 2010 Edition," USDA-ERS Economic Information Bulletin Number 66, 2010.

31. Steve Jennings et al., "Food in an urbanized world: the role of city region food systems in resilience and sustainable development," The International Sustainability Unit, The Prince of Wales Charitable Foundation, 2015.

32. Patricia Ballamingie et al., "Integrating a Food Systems Lens into Discussions of Urban Resilience: Analyzing the Policy Environment," *Journal of Agriculture, Food Systems, and Community Development*, 2020.

33. Local Governments for Sustainability (ICLEI) is the first and largest global network of local governments devoted to solving the world's most intractable sustainability challenges. The organization's standards, tools, and programs aim to credibly, transparently and robustly reduce greenhouse gas emissions, improve lives and livelihoods and protect natural resources in the communities they serve. iclei.org

34. The Milan Urban Food Policy Pact is an international agreement of mayors. More than a declaration, it is a concrete working tool for cities that includes a Framework for Action listing 37 recommended actions and indicators specific to each action to monitor progress towards implementing the Pact. The Milan Pact Awards offer concrete examples of the food policies that cities are implementing in each of the six Pact categories. milanurbanfoodpolicypact.org

35. Patricia Ballamingie et al., "Integrating a Food Systems Lens into Discussions of Urban Resilience: Analyzing the Policy Environment," *Journal of Agriculture, Food Systems, and Community Development*, 2020.

36. "Megaregions," America 2050, Regional Planning Association, 2015; "Knowledge Base: Food Systems," American Planning Association, 2021.

37. Michael Conard et al., "Infrastructure and Health: Modeling production, processing and distribution infrastructure for a resilient regional food system," Urban Design Institute, Columbia University, 2011.

38. Laura Lengnick et al., "Metropolitan Foodsheds: A Resilient Response to the Climate Change Challenge?" *Journal of Environmental Studies and Sciences*, 2015.

Chapter 14: The Way Forward

1. Laura Lengnick et al., "Metropolitan Foodsheds: A Resilient Response to the Climate Change Challenge?" *Journal of Environmental Studies and Sciences*, 2015.
2. You can get started learning more about city-led food system innovations for sustainability and resilience at "City Food Network," Local Governments for Sustainability (ICLEI); Milan Urban Food Policy Pact; "Scaling Up Climate Action: Food Systems," C40 Cities; and "Pilot Cities," City Region Food Systems Programme, UN Food and Agriculture Organization, 2021.

Chapter 15: Vegetables

1. Matt Grimley, "Farm Transitions: Klein & Breckbill," Land Stewardship Project, n.d.
2. Greg Padget, "From Land Protection to Land Access," Practical Farmers of Iowa, 2019.
3. Liz Kolbe, "Field Day Recap: Starting a Farm at Humble Hands Harvest," Practical Farmers of Iowa, 2018.
4. Singing Frogs Farm. Sebastopol, California. American Farmland Trust Stewardship Profiles, n.d.
5. Read more about Elizabeth Henderson and Peacework Farm at cultivating resilience.com.
6. To learn more about holistic farm management see Elizabeth Henderson and Karl North, *Whole-Farm Planning: Ecological Imperatives, Personal Values, and Economics*, Chelsea Green Publishing, 2011.
7. Read more about Mark Shepard in this volume.
8. Occultation, or "weeding with tarps," is considered an important weed management tool for biointensive no-till systems and is increasingly used in small-scale organic vegetable production. This practice typically involves irrigating a field or bed and laying down a thick, black plastic tarp for 2–8 weeks. To learn more see "Research and Recommendations on Occultation," Community Alliance with Family Farmers, 2020.
9. Living Energy Farm is an intentional off-the-grid community located in Louisa, Virginia, with the main goal of developing and promoting sustainability technology for the health of the planet. The Mission Statement of Living Energy Farm states: "The Living Energy Farm is a project to build a community, education center, and farm that demonstrates that a fulfilling life is possible without the use of any fossil fuel."

10. In 2016 Seed Savers Exchange in collaboration with Ira Wallace at Southern Exposure Seed Exchange requested over 60 varieties from the USDA to trial at Seed Savers Exchange, Heritage Farm in Decorah, Iowa. These varieties were collected by Edward H. Davis and John T. Morgan from seed savers across the Southeast, mostly in North and South Carolina. It is a priority of this project to revisit the original seed savers and learn about these collards' rich culinary and seed saving histories. "The Heirloom Collard Project," heirloomcollards .org

11. Callaloo (*Amaranthus viridis*) is a tasty, quick-growing, self-sowing hot weather green in the Amaranth family that is popular throughout the African diaspora, as well as in Asian cuisines. The abundant leaves are usually eaten cooked, and are sometimes referred to as Chinese spinach.

Chapter 16: Fruits and Nuts

1. *Camellia sinensis*.
2. *Ziziphus jujuba*.

Chapter 17: Grains

1. For a comparison of the certified organic and eco-farmed production practices used by Lundberg Family Farms' network of rice growers see "Frequently Asked Questions: Farming Practices," Lundberg Family Farms. lundberg.com /faq#farming-practices/

2. "Egg Salvage," California Waterfowl, 2020.

3. Because it is uniquely situated at the confluence of the Feather River, the Sacramento River and Butte Creek, Dos Rios is among the most frequently inundated floodplains in the Sacramento Valley. Heavy rains cause the rivers to back up and water to wash across the property. With the water come salmon, including the endangered winter run, that have free access to the floodplain (and all the nutrients it contains) as they migrate through the Sacramento River.

4. The no-till seeding of grains into pastures that have been mown or grazed before planting to reduce competition with the grain crop during establishment.

5. For another example of an early innovator of pasture-cropping in the U.S., see the Resilient Agriculture story about Tom Trantham's successful "12 Aprils" grazing program at his South Carolina pasture-based dairy in the first edition of *Resilient Agriculture* or online at cultivatingresilience.com

6. "USDA Task Force Clearing Up Cover Crop Rules," *No-Till Farmer*, 2013.

7. Ibid.

Chapter 18: Livestock

1. David Lewis, "Marin Carbon Project and the Carbon Connection," marin carbonproject.org

2. Shelby Vittek, "Seaweed May Be the Answer to the Burping Cow Problem," *Modern Farmer*, 2021.

3. The Cool Farm Alliance brings together farmers, NGOs, multinational food suppliers and retailers to promote agricultural practices that mitigate greenhouse gas emissions. Unilever, PepsiCo, Marks & Spencer, Tesco, Yara, Heineken and Fertilizer's Europe have joined forces as founding members of the Cool Farm Alliance whose mission is to enable millions of growers globally to make more informed on-farm decisions that reduce their environmental impact. Focusing on greenhouse gases in the first phase, the Alliance provides the Cool Farm Tool as a quantified decision support tool that is credible and standardized. The Cool Farm Alliance serves as both a knowledge platform and a tool development forum. coolfarmtool.org

4. *Consolidation and Concentration in the U.S. Dairy Industry*, Congressional Research Service Report, 2010.

5. Boyce Upholt, "A Killing Season," *The New Republic*, December 2018.

6. Drovers Road Preserve, droversroad.com

7. Jen Orris, "Next Generation: How Hickory Nut Gap Farm became a pioneer in grass-fed beef," Edible Asheville, nd.

8. savory.global/wp-content/uploads/2021/07/EOV-chapter-1-v3.pdf

9. savory.global/land-to-market/About the Author

Index

Page numbers in *italics* indicate tables and figures.

About the Author

LAURA LENGNICK is an award-winning soil scientist with 30 years of experience working as a researcher, policymaker, educator, activist and farmer to put sustainability values into action in U.S. food and farming. Her research in soil health and sustainable farming systems was nationally recognized with a USDA Secretary's Honor Award in 2002 and she served as a lead author of the 2013 USDA report, *Climate Change and Agriculture in the United States: Effects and Adaptation*. Laura is founder and owner of Cultivating Resilience, LLC, a private firm that works with organizations of all kinds to integrate resilience thinking into operations and strategic planning. In 2021, Laura joined the Glynwood Center for Regional Food and Farming as the Director of Agriculture. She resides in Asheville, NC.

Reading Guide available for download at:
https://newsociety.com/pages/resilient-agriculture-2-reading-guide

ABOUT NEW SOCIETY PUBLISHERS

New Society Publishers is an activist, solutions-oriented publisher focused on publishing books to build a more just and sustainable future. Our books offer tips, tools, and insights from leading experts in a wide range of areas.

We're proud to hold to the highest environmental and social standards of any publisher in North America. When you buy New Society books, you are part of the solution!

At New Society Publishers, we care deeply about *what* we publish—but also about *how* we do business.

- All our books are printed on 100% **post-consumer recycled paper**, processed chlorine-free, with low-VOC vegetable-based inks (since 2002). We print all our books in North America (never overseas)

- Our corporate structure is an innovative employee shareholder agreement, so we're one-third employee-owned (since 2015)

- We've created a Statement of Ethics (2021). The intent of this Statement is to act as a framework to guide our actions and facilitate feedback for continuous improvement of our work

- We're carbon-neutral (since 2006)

- We're certified as a B Corporation (since 2016)

- We're Signatories to the UN's Sustainable Development Goals (SDG) Publishers Compact (2020–2030, the Decade of Action)

To download our full catalog, sign up for our quarterly newsletter, and to learn more about New Society Publishers, please visit newsociety.com

ENVIRONMENTAL BENEFITS STATEMENT

New Society Publishers saved the following resources by printing the pages of this book on chlorine free paper made with 100% post-consumer waste.

TREES	WATER	ENERGY	SOLID WASTE	GREENHOUSE GASES
74 FULLY GROWN	**5,900** GALLONS	**31** MILLION BTUs	**250** POUNDS	**32,000** POUNDS

Environmental impact estimates were made using the Environmental Paper Network Paper Calculator 4.0. For more information visit www.papercalculator.org

23. The Urban Growers Collective is a Black- and women-led nonprofit farm in Chicago, Illinois, working to build a more just and equitable local food system. The collective cultivates eight urban farms on 11 acres of land, predominantly located on Chicago's South Side. These farms are production oriented but also offer opportunities for staff-led education, training and leadership development. The Growing Food Justice Initiative is a program of the Collective. urbangrowerscollective.org

24. For example, see this special issue of the *Journal of Agriculture, Food Systems and Community Development*: "The Impact of COVID-19 on the Food System," 2021; and coverage by popular media: Michael Corkery and David Yaffe-Bellany, "U.S. Food Supply Chain Is Strained as Virus Spreads," *New York Times*, April 13, 2020; Jill Colvin, "Trump orders meat processing plants to remain open," *AP News*, April 28, 2020; Victoria Knight, "Without Federal Protections, Farm Workers Risk Coronavirus Infection To Harvest Crops," *National Public Radio*, August 8, 2020; Brooke Jarvis, "The Scramble to Pluck 24 Billion Cherries in Eight Weeks," *New York Times Magazine*, August 12, 2020.

25. "There is No Future for Sustainable Farming Without Racial Justice," National Sustainable Agriculture Coalition, June 2020.

Chapter 13: Adding Resilience to the Menu

1. For example, Center for Whole Communities, *Whole Measures for Community Food Systems: Values-Based Planning and Evaluation*, 2009; "Infrastructure and Health: Modeling production, processing and distribution infrastructure for a resilient regional food system," Urban Design Lab at the Earth Institute, Columbia University, 2011; *Resilient Urban Food Systems: Opportunities, Challenges, and Solutions*, Resilient Cities Team, ICLEI-Local Governments for Sustainability, 2013.

2. See "The Adaptative Continuum" in Chapter 11.

3. Lauren Kelly, "Building Local Food Systems to Increase Resiliency," Institute for Sustainable Communities, n.d.

4. "Food and health: using the food system to challenge childhood obesity," Final Report on the Curbing Childhood Obesity Project, Phases I and II, Collaborative Initiatives at Massachusetts Institute of Technology, 2009.

5. Michael Conard et al., "Infrastructure and Health: Modeling production, processing and distribution infrastructure for a resilient regional food system," Urban Design Institute, Columbia University, 2011.

6. Christian Peters et al., "Foodshed Analysis and Its Relevance to Sustainability," *Renewable Agriculture and Food Systems*, 2009.

7. Walter Hedden, *How Great Cities Are Fed*, D. C. Heath, 1929.